T0231648

THE GEOLOGY OF THE BELINGWE GREENSTONE BELT, ZIMBABWE
A STUDY OF THE EVOLUTION OF ARCHAEAN CONTINENTAL CRUST

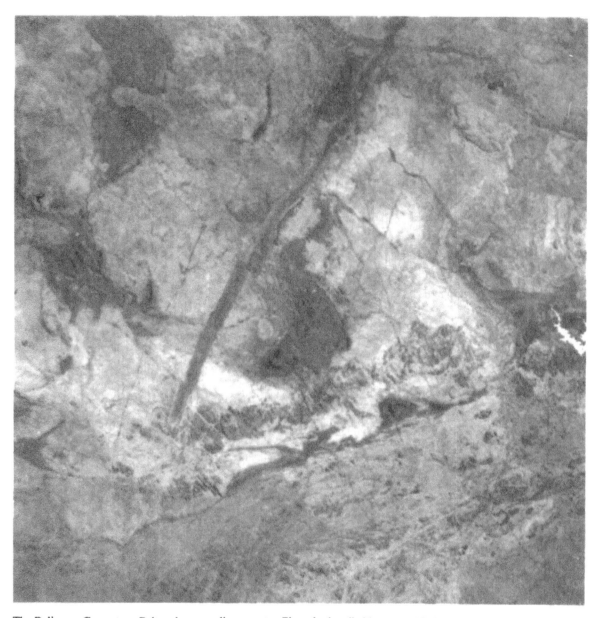

The Belingwe Greenstone Belt and surrounding country. Photo by handheld camera, NASA.

The Geology of the Belingwe Greenstone Belt, Zimbabwe

A study of the evolution of Archaean continental crust

Editors:

M.J. BICKLE
Department of Earth Sciences, University of Cambridge, Cambridge

E.G. NISBET
Department of Geology, Royal Holloway, University of London, Egham

Authors:

N.T. ARNDT, M.J. BICKLE, C. CHAUVEL, B. DUPRE, R.R. KEAYS,
A. MARTIN, E.G. NISBET, J.L. ORPEN & R. RENNER

GEOLOGICAL SOCIETY OF ZIMBABWE
SPECIAL PUBLICATION 2

CRC Press
Taylor & Francis Group
Boca Raton London New York

CRC Press is an imprint of the
Taylor & Francis Group, an **informa** business
A TAYLOR & FRANCIS BOOK

Published by
A.A. Balkema, P.O. Box 1675, 3000 BR Rotterdam, Netherlands
A.A. Balkema Publishers, Old Post Road, Brookfield, VT 05036, USA

ISBN 90 5410 120 2

Contents

V

Authors

N.T. Arndt, Institut de Géologie, Université de Rennes 1, Campus de Beaulieu, Avenue de Général Leclerc, 35042 Rennes Cedex, France

M.J. Bickle, Department of Earth Sciences, University of Cambridge, Downing Street, Cambridge CB2 3EQ, UK

C. Chauvel, Institut de Géologie, Université de Rennes 1, Campus de Beaulieu, Avenue de Général Leclerc, 35042 Rennes Cedex, France

B. Dupré, Institut de Physique du Globe, Université Pierre et Marie Curie, 4, place Jussieu, 75230 Paris Cedex 05, France

R.R. Keays, Department of Geology, University of Melbourne, Parkville, Victoria 3052, Australia

A. Martin, 6 Autumn Close, Greendale, Harare, Zimbabwe

E.G. Nisbet, Department of Geology, Royal Holloway, University of London, Egham, Surrey TW20 0EX, UK

J.L. Orpen, Geological Survey of Zimbabwe, P.O. Box 8039, Causeway, Harare, Zimbabwe

R. Renner, Department of Earth Sciences, University of Cambridge, Downing Street, Cambridge CB2 3EQ, UK

CHAPTER 1

Introduction: The Zimbabwe Craton and controversies over Archaean granite-greenstone terrains

M.J. BICKLE, E.G. NISBET, A. MARTIN & J.L. ORPEN

ABSTRACT

The Belingwe Greenstone Belt contains a wide variety of Archaean rocks which provide evidence of major significance in interpreting the stratigraphic history of the Zimbabwe Craton, and, more generally, in understanding the nature of the Archaean Earth. The stratigraphic sequences in the belt are well preserved, and the interpretation of the Belingwe rocks provides the underpinning of the present stratigraphic evolution of the Zimbabwe Craton. The sequence includes old basement, consisting of granitoid gneisses ca. 3.5 Ga old, overlain by three older greenstone sequences and a well developed younger (2.7 Ga) greenstone sequence. At the base of the 2.7 Ga sequence is a well exposed unconformity, demonstrating that the lavas and sediments of the younger greenstones were laid down on a mature terrain of eroded granitoid and greenstones.

The belt contains a wide variety of unusual rocks, including an extensive and locally very fresh suite of komatiites and komatiitic basalts, as well as carbonates which contain stromatolites. These rocks demonstrate that some greenstone belts formed on continental crust. The variety of sedimentary and volcanic environments probably reflects interaction of a range of tectonic processes as diverse as those operating today.

1.1 PURPOSE AND HISTORY OF PRESENT STUDY

The Belingwe Belt contains one of the most complete and best preserved greenstones sequences yet described. This study is a report of the results of several years (1973-1977, 1981-1983, 1986) of field investigation of the Archaean rocks within and around the Belingwe Greenstone Belt in Zimbabwe. The work was initiated as a field mapping program both to resolve basic geological relationships in the area and to search for insight into the more general problems of understanding the Archaean Earth. The mapping has been published by the Geological Survey of Zimbabwe (Martin 1978; Orpen et al. 1986). The stratigraphic relationships inferred in the area, which include a well-preserved basal unconformity to part of the supracrustal greenstone sequence, have led to a major revision of Archaean stratigraphy in Zimbabwe. The well preserved volcanic, sedimentary and fossiliferous rocks have provided much important petrological, geochemical and isotopic information. Here we report the field, petrological and geochemical data on which

1

interpretations of stratigraphy and field relations are based. The data are integrated into a tectonic model for the development of this block of early continental crust.

The work described here has its roots in the revolution in the Earth Sciences which took place in the late 1960's. Plate tectonics provided a coherent explanation of both the processes and fundamental causes of tectonic activity on the modern earth. Its success as an explanation of the workings of the modern Earth promoted enquiry into the evolution of tectonic processes over the earlier part of the Earth's history. Could detailed study of the earliest part of the geological record, the Archaean, provide a similar understanding of the young Earth? Two other factors added to the especial interest in Archaean geology at this time: 1) Improvements in radiometric methods of geochronology during the 1960's demonstrated the antiquity of Archaean rocks and provided a relative chronology for the first time; 2) magnesium-rich lavas or 'komatiites' were discovered in the Barberton Mountain Land by Viljoen & Viljoen (1969) and in Western Australia during the Nickel boom (e.g. Nesbitt 1971). The realisation that these ultramafic lavas must have erupted at high temperatures suggested that the thermal structure of the earth had changed substantially during earth history. This was the basis for numerous tectonic interpretations of Archaean geology. However the lack of consensus on these tectonic interpretations reflected fundamental problems which arose in part from uncertainty over the basic geological relationships within Archaean terrains. This study was initiated in order to gather much needed basic field information which could then serve as a foundation for interpretation.

Plate tectonic theory had illustrated the significant role of oceanic crust in modern tectonic processes and by the early 1970's it was realised how difficult it could be to distinguish allochthonous from autochthonous supracrustal sequences. A fundamental problem of Archaean geology was whether the greenstone sequences dominated by mafic volcanics were formed in oceanic environments and tectonically emplaced to their present crustal settings, or whether they had been laid down directly onto continental crust. Many authors in the early 1970's believed greenstone belts to be analogous to modern ophiolites: Many still do. The report that in the Belingwe Belt a basal unconformity existed between a greenstone belt succession and a granitoid-gneiss crust provided an unusual chance to evaluate the stratigraphic setting of some greenstone belt rocks. In addition to this basic aim, a study of well-preserved Archaean supracrustal rocks was likely to provide important information on Archaean igneous rocks, Archaean sedimentary environments and the geological setting of early life.

This study began in the early 1970's. In 1970 and 1971 M.J. Bickle and E.G. Nisbet became interested in the contrasts and similarities between young ophiolite complexes and Archaean ultramafic bodies. Separately, A. Martin mapped the northern half of the Belingwe Greenstone Belt for the Zimbabwe (formerly Rhodesian) Geological Survey during 1973, 1974 and 1975 and concurrently began doctoral research under the supervision of J.F. Wilson. In 1974 M.J. Bickle, on a postdoctoral research fellowship at the University of Zimbabwe (formerly University of Rhodesia) commenced work in the south-eastern part of the greenstone belt, and was joined by E.G. Nisbet on a NERC postdoctoral fellowship held at Oxford. This field work continued through 1975 with a short field seasons in 1976 (M.J.B.), 1982 (E.G.N.), 1983 and 1984 (A.M., J.L.O.). and further mapping and drilling in 1986 (E.G.N., M.J.B., A.M.). J.L. Orpen studied the south-west part of the greenstone belt in an undergraduate project in 1974 and undertook doctoral work in the same area from 1975 to 1977 supervised by J.F. Wilson. The field studies were curtailed by escalation of hostilities in Zimbabwe in 1976. For this reason the study is not as complete in some areas as the authors originally intended. The field studies were complemented by a number of laboratory studies. Major and trace element

2

concentrations of volcanic rocks were analysed in Oxford, Zurich and Leeds and more recently in Mainz and Saskatoon. The Rb-Sr isotopic studies on volcanic and granitoid rocks from the area were undertaken in Oxford and Leeds in conjunction with C.J. Hawkesworth, S. Moorbath, P. N. Taylor, P.J. Hamilton and R.K.O'Nions. Experimental studies on komatiites were undertaken initially at Edinburgh in conjunction with C.Ford and later at Lamont with D. Walker. Stable and radiogenic isotope and rare gas studies have been carried out at Saskatoon (by T.K. Kyser and E. Hegner) and at the University of Rhode Island (with P. Abell). Many of the laboratory studies on rocks from the Belingwe Greenstone Belt are still continuing especially in Rennes, Cambridge and Saskatoon and samples have been provided to a wide variety of other laboratories. A bibliography of work on Belingwe is listed at the end of the references. It should be noted that much of the initial inspiration and encouragement to work in the area for all the authors came from J.F. Wilson of the University of Zimbabwe who took a continuing interest in the field studies and their regional significance to the Zimbabwe Archaean. We hope that this work will be seen as initially envisaged: As a continuation of the tradition of field investigation established in Zimbabwe by F.P. Mennell and A.M. Macgregor.

1.2 ARCHAEAN GEOLOGY – SOME OF THE PROBLEMS

The Archaean record extends over one third of earth history from ca. 4.3 or 4.2 Ga ago until approximately 2.5 Ga, when the Great Dyke was emplaced in Zimbabwe. The Archaean thus provides most of the available data about the critical stages in the early evolution of the earth. Our ability to interpret this geological record is hindered by some very special problems. Tectonic and surface geological processes, may have differed substantially from modern processes, making uniformitarianism suspect. Preservation of terrains is likely to have been selective and the preserved sample may be nonrepresentative. The size of most Archaean terrains is relatively small. Some are comparable to modern orogenic provinces, others are mere fragments: This makes resolution of tectonic provinces difficult. Fossils are absent or not stratigraphically useful, and chronostratigraphy is possible only where U-Pb zircon geochronology is available. In most Archaean terrains it is difficult to resolve even relatively major geological events. For these reasons it is perhaps not surprising that there is little consensus in current interpretations of Archaean geology and the interpretations are invariably simplistic. An additional problem is that many Archaean terrains are in relatively remote or inaccessible and less populated parts of the world and have not been mapped in the detail to which more recent orogenic belts are known.

Archaean terrains are conveniently grouped into the high-grade gneiss terrains, granite-greenstone terrains and late Archaean sediment-dominated cratonic basins. The relationships of these three environments are not properly known. It should be noted that cratonic sequences such as the 3000 Ma Pongola and 2700 Ma (?) Witwatersrand successions in S. Africa and 2700 Ma Fortescue Group in the Pilbara, W. Australia were being deposited on relatively rigid cratons over the same period as much of the Belingwe Greenstone Belt was laid down. Similarly, the Limpopo Belt, a diverse orogenic belt including high grade gneiss, which lies only 50 km south-east of Belingwe, comprises rocks ranging from early to late Archaean in age which underwent major tectono-metamorphic events at 3300 and 2900 Ma (Hickman & Wakefield 1975). These events to some extent were synchronous with the development of the granite greenstone terrain of the Zimbabwe Craton.

The granite-greenstone terrains are the most distinctive type of Archaean outcrop and

3

the mafic-volcanic dominated supracrustal greenstone sequence disposed around ovoid or more linear granitoid and gneiss batholiths exhibit a tectonic style not often seen in younger provinces. Since Macgregor (1951) drew attention to the distinctive pattern of 'gregarious' batholiths there has been much speculation on the origin of granite-greenstone terrains. This speculation has drawn attention to a number of unresolved geological relationships within the terrains as discussed by Macgregor. First, there is the problem of the nature of the contact between the basement and the supracrustal greenstone sequences. Within Archaean terrains most granite-greenstone contacts are either intrusive or tectonic. Macgregor described important within-greenstone unconformities in Zimbabwe and speculated (1947) that the Belingwe Greenstone Belt was unconformable on a granite-gneiss basement. This unconformity was later proved by Laubscher (1963) but not described in detail until the present work (Bickle et al. 1975). From an understanding of the regional field geology Macgregor inferred an older granitoid gneiss and greenstone basement to the main greenstone sequences.

Outside Zimbabwe, the lack of any clear unconformities in many granite-greenstone areas and the advent of plate-tectonic interpretations – with the recognition of the significance of allochthonous oceanic rocks in orogenic terrains (ophiolites) – led to alternative interpretations of greenstone sequences as disrupted allochthonous oceanic crust (e.g. Anhaeusser 1973; Burke et al. 1976; Goodwin & Ridler 1970). Distinction between the 'autochthonous continental' or the 'allochthonous oceanic' models is a crucial initial step in tectonic interpretation of granite-greenstone terrains. Are greenstone sequences typically disordered, allochthonous and bounded by thrusts or are they stratigraphically intact autochthonous sequences? This problem (which may have different answers in different places) can only be resolved by direct observation of the evidence for tectonic, or stratigraphic and unconformable relations, observation of local sedimentological environments or larger scale stratigraphic correlation and sedimentological facies variation. This was an important aim of our study. In addition, there are the related questions about the original extent of greenstone sequences and the possibility of regional stratigraphic correlation. The presently preserved greenstone belts may be remnants from a once continuous cover or may have been individual basins, filled with sediment and volcanic rocks, with dimensions similar to the present size of the belts.

It is perhaps a measure of the primitive state of Archaean geological interpretations that few if any general subdivisions of the tectonic environments in which greenstone belts formed are widely recognised or accepted. Greenstone belts contain most of the rock types found in Phanerozoic successions and exhibit significant variations in rock content, tectonic style and metamorphic grade. It might be expected that these supracrustal sequences were formed in as diverse a range of tectonic settings as is seen in Phanerozoic sediment-volcanic associations. The few attempts to establish either chronological variations in greenstone belt style (e.g. Glikson 1979) or spatial variations in greenstone belt environments (e.g. Groves & Batt 1984) remain unconvincing because the environmental interpretations on which they are based are poorly documented.

The sedimentary and volcanic rocks of the greenstone sequences also contain information on two other aspects of Earth evolution. Volcanic rocks provide a 'window' to mantle processes and mantle composition. Komatiites, which are magnesium-rich lavas erupted at high temperature, are mostly restricted to the Archaean. They form a minor but distinctive part of many greenstone sequences and have attracted much attention because study of these rocks has produced geochemical and thermal constraints on the Archaean mantle. A key problem is understanding the thermal history of the Earth. In the Archaean, there was higher radiogenic heat production in the mantle: Was the mantle hotter? Komatiites provide direct evidence of higher mantle temperatures. Sedimentary rocks reflect the composition of the atmosphere and oceans and Archaean sediments provide important information on the early evolution of the hydrosphere. Archaean rocks contain the oldest fossils and provide sparse clues towards the timing and origin of life. The impact of life on the composition of the Archaean atmosphere, especially the partial pressure of oxygen, is a topic of great interest.

Another important area of controversy concerns the structure of the upper crust, and in particular the significance of the 'gregarious' batholith pattern distinctive of Archaean granite-greenstone terrains. Neither mafic-dominated supracrustal sequences nor granitoid batholiths per se are restricted to the Archaean. What makes the Archaean terrains distinctive is the two-dimensional outcrop of batholiths and greenstones over thousands of square kilometres. The controversy on the origin of the batholiths has paralleled the controversy on the origin of younger gneiss domes (mantled gneiss domes of Eskola, 1949). There are two main hypotheses: 1) The domal pattern is a consequence of externally imposed events such as cross-folding or gravitational instability on a tectonically thickened crust; or 2) the batholiths are internally generated and their emplacement is the cause of sedimentation and then the deformation and metamorphism in the adjacent greenstone belts. The relationship (if any) between uprise of the granitic batholiths and sedimentation and the deformational patterns in the greenstones and granitoid crust should provide tests capable of distinguishing these two hypotheses.

Syntheses based either on observations of Archaean rocks or more directly thermal/tectonic models of earth evolution will undoubtedly play a part in understanding processes in the Archaean earth. However, until very detailed geochronological control is available and much more is known about geological and especially tectonic, stratigraphical, sedimentological and volcanological relationships over 10 to 100 s of kilometres, we question whether it will be possible to make realistic regional tectonic syntheses. We hope to demonstrate that the results of this study of the Belingwe Belt, admittedly incomplete, indicate how much is still to be learnt by careful basic field geology and that application of methods of investigation familiar in Phanerozoic orogenic belts can be successful in Archaean terrains.

1.3 OUTLINE OF THE REGIONAL GEOLOGY

The Belingwe Greenstone Belt is located in the region around the towns of Mberengwa and Zvishavane (Fig. 1.1) in the south-east of the Zimbabwe Archaean Craton which forms the central part of Zimbabwe and extends into Botswana to the south-east and Mozambique to the north-east. In this memoir we reserve the term 'craton' to describe that area of low grade granite-greenstone terrain which has undergone little deformation since the close of the Archaean. We place no genetic implications on the term.

Around its western, northern and eastern margins the Zimbabwe Craton is either covered by younger strata or abuts younger orogenic provinces. On south-eastern and south-western margins of the craton is the Limpopo Belt which is composed of high grade gneissic rocks 3600 to 2600 Ma in age. This age range is similar to that of rocks on the craton. Despite the original concept of Archaean cratons as little deformed relics of old crust surrounded by younger mobile belts (Clifford 1970) it is now apparent (Barton & Key 1981; Taylor et al. 1991) that the craton and the Limpopo Belt both record complex tectonic histories over most of the Archaean. The present contact between these units is in some areas tectonic with Limpopo Belt gneisses in late thrust contact on 'cratonic' granitoids; elsewhere a gradational increase in strain occurs towards the Limpopo Belt (Coward et al. 1976).

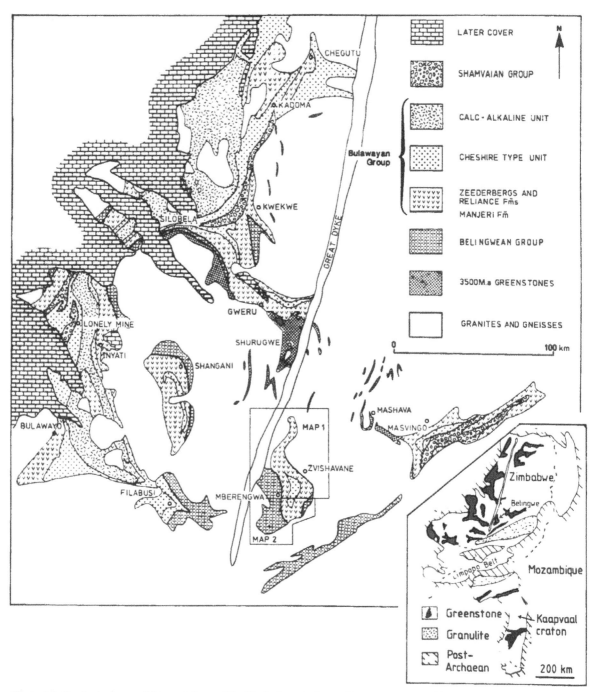

Figure 1.1. Geological map of the central part of the Zimbabwe Archaean Craton. Maps 1 and 2 on diagram refer to Geological Survey maps. Map 1 is Zimbabwe Rhodesia Geological Survey Map, Belingwe-Shabani, 1:100 000, Martin (1978). Map 2 is Zimbabwe Geological Survey Map, 1:100 000, Belingwe Peak, Orpen et al. (1986). These maps are available from the Geological Survey of Zimbabwe, P.O. Box 8039, Causeway, Zimbabwe.

The craton itself consists of varied granitic and gneissic rocks ranging in age from 3600 to 2500 Ma (Wilson et al. 1978; Wilson 1979; Moorbath et al. 1977; Taylor et al. 1991), with widespread greenstone belts containing an assortment of rocks of similar age range. The craton is cut by the Great Dyke (2460 ±16 Ma, Rb-Sr mineral isochron, Hamilton 1977) and by dykes of Mashonaland dolerite (1800 Ma). The subdivision of the craton into greenstone belts ('gold belts') and granitoids was early recognised (Fletcher & Espin 1897; Mennell 1904). Subsequent work based on detailed geological survey mapping at 1:100 000 has been concerned mainly with attempting to comprehend the stratigraphy of the greenstones and more recently into subdividing the granitoids. Macgregor (1947) proposed the first major subdivision. He recognised a lower ultra-mafic dominated unit, the Sebakwian, overlain unconformably by a unit dominated by mafic volcanic rocks, the Bulawayan, and above it an unconformable upper sedimentary unit, the Shamvaian. This 'simple' triad stratigraphy survived essentially unchanged for nearly thirty years. In the 1970's, geochronological work (initially K-Ar; Wilson & Harrison 1973) and more recently Rb-Sr, Sm-Nd, Pb-Pb whole-rock and U-Pb zircon methods (Hawkesworth et al. 1975; Hamilton et al. 1977; Hickman 1978; Moorbath et al. 1977; Hawkesworth et al. 1979; Moorbath et al. 1987; Taylor et al. 1984, 1991; Dodson et al. 1988) demonstrated the great age range of greenstone and granitoid rocks on the craton. The stratigraphic relationships described in this memoir (Table 1.1) have prompted a craton-wide revision of the stratigraphy, since it is possible to correlate the Belingwe succession over much of Zimbabwe (Wilson et al. 1978; Wilson 1979, 1981). This correlation has produced a stratigraphy in general compatible with but rather more complex than that proposed by Macgregor (Stagman 1978).

As discussed in more detail in the following chapters we now recognise three main greenstone sequences with ages of ca. 3500 Ma, ca. 2900-2800 Ma and 2700 Ma. The ages of the two older sequences are uncertain, being inferred from indirect evidence as discussed below. The greenstone sequences are preserved in a wide range of tectonic states. Much of the youngest sequence is preserved in what appear to be synclinal belts with relatively little internal strain (except in local shear zones) and with metamorphic recrystallisation only to greenschist facies. Locally, especially near granitoid contacts, higher strains and metamorphic grade convert the supracrustal rocks to foliated amphibolite-facies gneisses. Only in the south-west of Zimbabwe is there evidence for major overturning of this 2.7 Ga supracrustal sequence in nappe-like structures (Coward et al. 1976). The 2900 Ma greenstones exhibit a range in tectonic setting which is similar to the style of the unconformably overlying and more widespread 2700 Ma greenstones. The 3500 Ma greenstones are best preserved on the overturned limb of a major nappe structure at Shurugwe (formerly Selukwe; Cotterill 1969) and more commonly as thin, high strain, foliated amphibolite facies intercalations within gneisses of similar age. The granitoid crust is also very heterogeneous. High strain gneissic rocks of 3500 Ma age contain areas of much less foliated or cross-cutting plutonic rocks as at Zvishavane (e.g. Moor-

Table 1.1. Regional stratigraphy: Archaean of central Zimbabwe.

Old usage (Stagman 1978)		1970's usage	Belingwe stratigraphy
	Shamvaian group	Shamvaian group	
Bulawayan group	Upper greenstones	Upper Bulawayan	Ngezi group
	Lower greenstones	Lower Bulawayan	Mtshingwe group
Sebakwian group	Schist inclusions	Sebakwian group	

bath et al. 1977; Taylor et al. 1991) and in the Ngezi schist inclusion (Chapter 3). Elsewhere high strain 2900 Ma gneisses are intruded by little deformed 2900 Ma granitoids (Hawkesworth et al. 1979; Taylor et al. 1991). The distribution and timing of high strain events within the granitoid crust is not well known. In general the younger plutonic rocks, particularly the widespread 2700 Ma tonalites (Taylor et al. 1991) and 2570 Ma adamellite (Wilson et al. 1978; Wilson 1979) which intrude all ages of greenstone, are much less intensely deformed.

1.4 SETTING OF THE BELINGWE GREENSTONE BELT

The Belingwe Belt is situated between gneissic granitoid crust of 3500 Ma age to the east and 2900 Ma age to the west. It contains a lower (possibly 2900 Ma) sequence of greenstones collectively termed the Mtshingwe Group which is unconformably overlain by another, later, sequence of greenstones, the Ngezi Group. Foliated supracrustal relicts are found intercalated with both the 3500 Ma gneisses and the 2900 Ma gneisses. Both the Mtshingwe Group and Ngezi Group greenstones have what appear to be relatively simple synclinal dispositions and contain a wide range of sedimentary and volcanic rock type. These rocks often show little internal strain and are in the main only metamorphosed to greenschist facies or less, except close to younger granitoid gneisses. The proposal that the Bulawayan Supergroup be restricted to the Ngezi Group and craton-wide correlatives, and that the term 'Belingwean' Supergroup be introduced to cover the Mtshingwe Group and correlatives, is discussed below, in Chapter 2.

1.5 PREVIOUS WORK ON THE BELINGWE GREENSTONE BELT

Early work in the area included regional reconnaissance by Fletcher & Espin (1897) who outlined the Belingwe Greenstone Belt with surprising accuracy on their geological map of Matabeleland, and the work of Mennell (1904, 1905). The first detailed mapping was by Keep (1929). This work included the Shabani ultramafic complex and an adjacent segment of the northern part of the greenstone belt. Keep also described in detail so-called 'Limburgite' greenstones, now recognised as komatiite lavas, and identified by him as ultrabasic. Worst (1956) produced a 1:100 000 map of the whole southern portion of the belt. Laubscher (1963) described the Shabani complex in detail and discovered and mentioned the unconformity between the basal sediments and the gneisses on the northeast margin of the belt. Macgregor (1951) had previously suggested that this contact might be an unconformity, on the basis of the mapping by Keep (1929) and the low metamorphic grade of the basal sediments which strike parallel to the contact, in contrast to the higher grade gneiss discordant to the contact. Keep had interpreted the contact as intrusive. The unconformity outcrop was first described in a field trip guide, as part of the 1971 granite conference of the Rhodesian branch of the Geological Society of South Africa (Morrison & Wilson 1971). Oldham (1968, 1970) compiled a map of almost the whole northern area and also mentioned the unconformity. Wilson (1973) cited the unconformable basal relations as evidence for a continental setting for the Belingwe Greenstone Belt.

The first full descriptions of the basal unconformity, recognition of komatiitic volcanic rocks and discovery of stromatolitic limestones were made in this study. Amongst the various publications (which are listed in the bibliography at the end of the references) on the topic are: Bickle et al. (1975) describing the unconformity, the basic stratigraphic relationships, the geochemistry of the volcanics and the stromatolites; Nisbet et al. (1977), a detailed description of the geology, petrology and geochemistry of the volcanic

8

rocks; Martin (1978), a geological survey bulletin and map describing the northern half of the greenstone belt; Wilson et al. (1978) and Wilson (1979) which are reviews of Zimbabwe stratigraphy in the light of the Belingwe stratigraphy; Orpen's (1978) Ph.D. thesis on the geology of the south-western part of the belt; Martin et al. (1980), a detailed description of the stromatolite localities; Martin's (1983) Ph.D. thesis on the northern half of the belt; Abell et al. (1985a, b) on the stromatolites; Orpen et al.'s map (1986) of the southern half of the belt; and Nisbet et al. (1987) on fresh komatiites from the area near Zvishavane.

At the same time as this basic geological description was published, several geo-chronological studies of both greenstones and granitoid in the Rhodesian (now Zimbabwe) Craton relevant to the Belingwe area were undertaken. Hawkesworth et al. (1975), Jahn et al. (1976), Hamilton et al. (1977) and Taylor et al. (1991) made regional studies of the greenstone ages. Detailed Sm-Nd and Pb-Pb whole-rock studies by Chauvel et al. (1983) of the younger greenstones (Ngezi Group) at Belingwe now give a reliable age of 2690 ± 13 Ma for eruption of these rocks. Dodson et al. (1988) have dated detrital zircons from the Wanderer Formation, Shurugwi, and from the Buchwa quartzite, Mweza Greenstone Belt. Both units contain zircons as old as 3800 Ma, testimony to crustal sources older than any rocks yet dated on the craton. Hickman (1974), Hawkesworth et al. (1975), Hawkesworth & Bickle (1977), Moorbath et al. (1977), Hickman (1978), Hawkesworth et al. (1979) and Taylor et al. (1991) carried out whole-rock Rb-Sr, Pb-Pb and Sm-Nd analyses of granitoid rocks from the vicinity of the Belingwe Greenstone Belt and distinguished the three main groups of plutonic intrusion at 3500 Ma, 2900 Ma and 2600 Ma. The geochronological work has greatly improved the interpretation of the geology of the Belingwe area but much is still unknown. The most important problem is that the age of the Mtshingwe Group greenstones, which are unconformably overlain by the 2.7 Ga Ngezi Group greenstone, remains poorly constrained. Also many of the ages are relatively imprecise compared with the detailed relative chronology of the stratigraphy now known and the majority of the ages are based on Rb-Sr isochrons which are liable to be perturbed in Archaean crust with a complex history (e.g. Bickle et al. 1983; Taylor et al. 1991).

1.6 ECONOMIC SIGNIFICANCE TO ZIMBABWE OF THE BELINGWE BELT

The stratigraphic framework discussed above is a fundamental guide to exploration for economic mineral deposits (e.g. Wilson 1979). Thus nickel deposits in Zimbabwe are found in correlatives of the Reliance Formation and a number of gold deposits are localised in Manjeri Formation and on the contact between the Zeederbergs and Cheshire Formations (Nisbet 1983). Limestones, some of which are stromatolitic, are restricted to the Manjeri and Cheshire Formations.

1.7 CONVENTIONS ADOPTED

1.7.1 *Names*

New place names are used throughout the text. In conformance with stratigraphic law greenstone belts retain their original names. Since the old place names are extensively used in the literature equivalents are listed below.

9

Belingwe	Mberengwa	Lundi	Runde
Fort Victoria	Masvingo	Shabani	Zvishavane
Gwelo	Gweru	Selukwe	Shurugwe
Que Que	Kwekwe	Umvuma	Mvuma
Gatooma	Kadoma	Mashaba	Mashava
Hartley	Chegutu	Enkeldoorn	Chivhu
Mtshingwe River		Salisbury	Harare
(also Mtschingwe R.)		Umtshingwe River	

1.7.2 *Rock terminology*

Despite the ubiquitous low grade metamorphism we choose to describe originally sedimentary and volcanic rocks by their original character, not by their metamorphic assemblages. This is because much of the interesting information available in the rocks pertains to their original character, and preservation is for the most part adequate to identify without doubt original features. The terminology of rocks in the komatiite suite is from Arndt & Nisbet (1982), and the classification of granitic rocks is as proposed by Streckeisen (1976).

We have continued the usage of 'banded ironstone' to describe laminated chert-iron rocks rather than the recommended 'iron-formation' (James 1951). This is to avoid such linguistic tongue twisters as 'Bend Formation iron-formation.' Quoted Rb-Sr isotopic ages have been recalculated with the decay constants of Steiger & Jaeger (1977) where necessary and all errors are quoted at 2σ.

ACKNOWLEDGEMENTS

This project was based on the commitment of the then Geological Survey of Rhodesia (now Zimbabwe) and the University of Rhodesia (Zimbabwe) to continued field research. Mr. F. P. Tennick suggested that A. Martin map the northern part of the Belingwe area for the Geological Survey under the direction of Dr. J.W. Wiles and later Dr. J.G. Stagman. Prof. G. Bond and Prof. J.F. Wilson played a crucial role in discussions with M.J.B. and E.G.N. on an initial choice of field area and the late Professor Bond went to considerable lengths to provide field support. The painstaking introduction to and continuing discussions on Zimbabwe geology provided by Prof. J.F. Wilson were especially appreciated by all the authors. In addition Prof. J.F. Wilson supervised doctoral studies by A.M. and J.L.O. on the area. Technical support, especially that of J. Morgan, from the University of Zimbabwe is acknowledged. Sources of financial support included postdoctoral research fellowships from the University of Zimbabwe (M.J.B.), N.E.R.C. (M.J.B. & E.G.N.), Geological Survey of Zimbabwe (A.M.), a University of Zimbabwe Research Scholarship (J.L.O.), the Royal Society European exchange program, the Swiss Nationalfond and Wolfson College, Oxford (E.G.N.), as well as the Natural Sciences and Engineering Research Council of Canada. The Universities of Oxford, Leeds, Cambridge, Western Australia, E.T.H.-Zurich and Saskatchewan, the Max Planck Institute and the University of Zimbabwe have provided facilities to help the project, which began as collaboration between Oxford University and the University of Rhodesia.

We would like to thank local farmers for access and assistance especially Mr. Bill Smith, Mr. M. Walker and Mr. A. Rauch at Texas Land and Cattle Co. (Union Carbide Zimbabwe), Mr. V. Cockroft, Mr. Moorcroft, Mr. P. Hall and Senator Garfield Todd. Mnene mission provided much welcome and kind hospitality: We are especially grateful to the staff there.

We would like to acknowledge fruitful discussions with all the visitors to the area during the period of the work including Drs. Sutton, Coward, Shackleton, Oldham, Fripp, Williams, Preston Cloud, Oxburgh, de Wit, Hawkesworth, Button, Maiden, Barton, Arndt, Keays, Condie, Cameron, Cotterill, Coleman, Droop, Holland, Harris, Renner, Cheadle, Scholey, and especially J.F. Wilson.

Probably the worst task fell to our critical readers and those involved in publishing the work, who ploughed resolutely through the document. We thank them all, and especially Mike Coward, Ken Eriksson, Jan Kramers, Dan McKenzie, Ron Oxburgh and especially Stephen Moorbath for encouragement and helpful criticism.

Support by NSERC Canada and various University of Saskatchewan funds was vital to the final compilation of this memoir in allowing M.J. Bickle and A. Martin to visit Canada. The manuscript was typed, with great patience, by U. Stidwill, S. Ford and, mostly, by Angie Heppner. Hilary Alberti drew many of the figures.

11

CHAPTER 2

Rock units and stratigraphy of the Belingwe Greenstone Belt: The complexity of the tectonic setting

A. MARTIN, E.G. NISBET, M.J. BICKLE & J.L. ORPEN

ABSTRACT

The Belingwe Greenstone Belt, in south-east Zimbabwe, contains a well-preserved succession of Archaean lavas and sedimentary rocks, laid down on older continental crust. Surrounding the belt is a varied granite-gneiss terrain ranging from 3.6 to 2.6 Ga in age. The supracrustal rocks consist of two distinct sequences. The lower sequence, the Mtshingwe Group, includes the Hokonui (conglomerates and acid volcanic rocks), Bend (komatiitic lavas and ironstones), Koodoovale (conglomerates and acid volcanic rocks) and Brooklands (komatiitic lavas and varied sedimentary rocks) Formations. The sequence was probably laid down on a basement of older gneiss. Lying unconformably above both the Mtshingwe Group and the older granite-gneiss terrain, with well exposed basal contacts, is the 2.7 Ga Ngezi Group. This consists of the Manjeri (thin sedimentary rocks), Reliance (komatiites), Zeederbergs (basalts) and Cheshire (mainly shallow water sedimentary rocks) Formations. The Manjeri and Cheshire Formations both contain stromatolitic horizons.

The stratigraphy of the Zimbabwe Archaean Craton has been reassessed in the light of work in Belingwe and it has been shown (Wilson et al. 1978; Wilson 1979) that the Belingwe succession serves as a basis for a craton-wide correlation.

2.1 INTRODUCTION

The Belingwe Greenstone Belt contains a variety of igneous and metamorphic rocks, set in a granite-gneiss terrain. Five main tectonic units are identified (Fig. 2.1). The oldest unit is the ca. 3.5 Ga Shabani-Tokwe Gneiss complex, forming much of the terrain to the east of the Belingwe Belt and partly underlying much of the greenstones. The second unit consists of ca. 2800 - 2900 Ma gneisses and granites, including the Mashaba Tonalite east of Belingwe and the Chingezi Gneiss and Chingezi Tonalite to the west.

Two distinct greenstone sequences occur. The older, lower greenstones, collectively termed the Mtshingwe Group, comprise the Hokonui, Bend, Brooklands and Koodoovale Formations; the younger upper greenstones, the Ngezi Group, comprise the Manjeri, Reliance, Zeederbergs and Cheshire Formations (Table 2.1).

On the east side of the belt the basal unit of the upper Ngezi Group, the Manjeri Formation, lies unconformably on the Shabani Gneiss; in the south and west of the belt

13

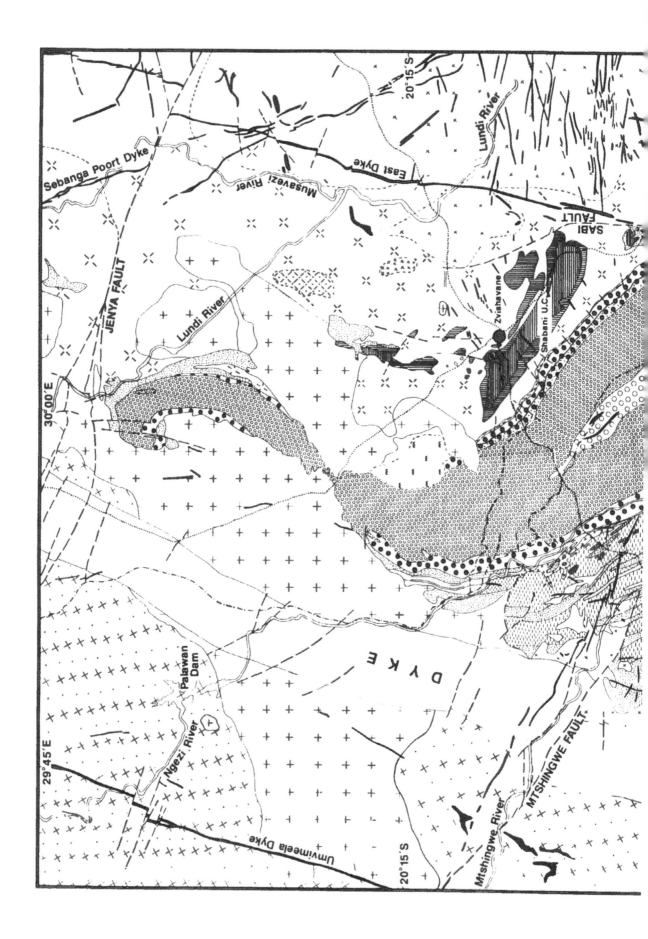

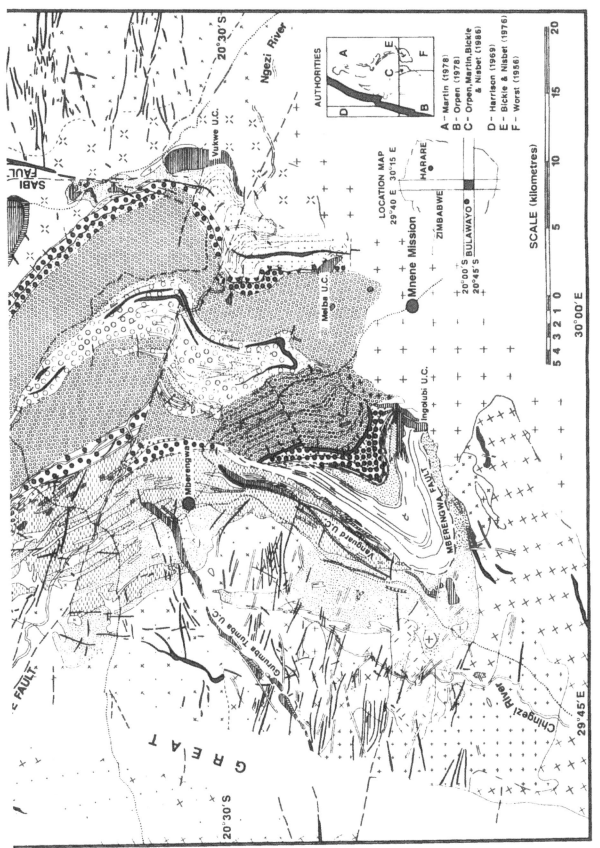

Figure 2.1. Distribution of principal geological units in the area of the Belingwe Greenstone Belt.

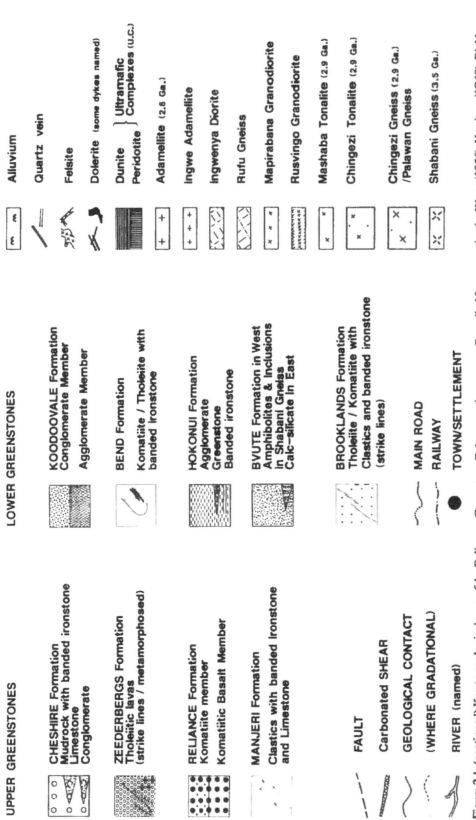

UPPER GREENSTONES

CHESHIRE Formation
Mudrock with banded ironstone
Limestone
Conglomerate

ZEEDERBERGS Formation
Tholeiitic lavas
(strike lines / metamorphosed)

RELIANCE Formation
Komatiite member
Komatiitic Basalt Member

MANJERI Formation
Clastics with banded Ironstone
and Limestone

FAULT

Carbonated SHEAR

GEOLOGICAL CONTACT

(WHERE GRADATIONAL)

RIVER (named)

LOWER GREENSTONES

KOODOOVALE Formation
Conglomerate Member
Agglomerate Member

BEND Formation
Komatiite / Tholeiite with
banded Ironstone

HOKONUI Formation
Agglomerate
Greenstone
Banded Ironstone

BVUTE Formation in West
Amphibolites & Inclusions
in Shabani Gneiss
Calc–silicate in East

BROOKLANDS Formation
Tholeiite / Komatiite with
Clastics and banded ironstone
(strike lines)

MAIN ROAD

RAILWAY

TOWN/SETTLEMENT

Alluvium

Quartz vein

Felsite

Dolerite (some dykes named)

Dunite ⎱ Ultramafic
Peridotite ⎰ Complexes (u.c.)

Adamellite (2.6 Ga.)

Ingwe Adamellite

Ingwenya Diorite

Rufu Gneiss

Mapirabana Granodiorite

Rusvingo Granodiorite

Mashaba Tonalite (2.9 Ga.)

Chingezi Tonalite (2.9 Ga.)

Chingezi Gneiss (2.9 Ga.)
/Palawan Gneiss

Shabani Gneiss (3.5 Ga.)

Figure 2.1. (continued).Key to geological map of the Belingwe Greenstone Belt on previous pages. Compiled from mapping by Worst (1956), Harrison (1969), Bickle & Nisbet (1976), Martin (1978), Orpen (1978) and Orpen et al. (1986).

Table 2.1. Major geological units: Belingwe area.

Intrusive units	Supracrustal units		Lithology
	Groups	Formations	
Great Dyke 2460 ± 60 Ma Chilimanzi Suite-Chibi Batholith 2570 ± 25 Ma			
		Cheshire	Conglomerate, 1st, shale
	Ngezi Group 2692±9	Zeederbergs	Basalt
		Reliance	Komatiite, komatiitic basalt basalt
		Manjeri	clastic sediments
	Unconformity		
Mashaba Tonalite 2870±160 Ma Chingezi Tonalite 2833±43 Ma			
		Koodoovale	Conglomerate, felsic agglomerate
		Bend	Komatiite, komatiitic basalt, ironstone
		Brooklands*	Conglomerate, siltsone, quartzite, ironstone, komatiite, komatiitic and tholeiitic basalt
		Hokonui	Intermediate to felsic volcanics, mostly pyroclastic, mafic volcanics
	Possible unconformity		
		Bvute**	Amphibolite
Chingezi Gneiss 2810 ± 70 Ma			Banded grey gneisses
Shabani Gneiss Complex ca. 3500 Ma			Banded grey gneisses 'Schist inclusions'

*Correlation of Brooklands Formation in east with other Mtshingwe Group formations in west is not known. **Stratigraphic and structural relationships of Hokonui and Bvute Formations and Chingezi Gneiss are uncertain.

this unconformity transgresses onto the Mtshingwe Group. It is probable that the Brooklands Formation of the Mtshingwe Group also lies unconformably on gneisses of the Shabani Gneiss Complex. The basal contact of the Mtshingwe Group on the western side of the greenstone belt is not well understood and is discussed further below.

Younger than all the greenstones are the granite plutons of the Chilimanzi Suite, and the Great Dyke, as well as various other suites of dykes.

In the following discussion we attempt to provide a stratigraphic framework to the various separate publications on the Belingwe Belt. The structure and geochronology of the region is described, with particular reference to the gneissic and granitic rocks surrounding the belt. Then a stratigraphic outline of the Ngezi and Mtshingwe Groups is given.

Figure 2.2. LANDSAT image, Belingwe Greenstone Belt (lower right), Great Dyke (centre). Note that the most striking dark/light contrasts, bounded by straight lines, are land-use patterns distinguishing communal regions without trees (light) from commercial farms with trees (dark). Great Dyke runs N-S through centre of image, Belingwe Belt is dark region in centre, Limpopo Belt is in S-W corner, with Chibi batholith running SW-NE between Belingwe Belt and Limpopo Belt. Compare with Figure 1.1.

Table 2.2. The main tectonic units.

Younger (Chibi) Granites	2.6 Ga
Ngezi Group	2.7 Ga
Mtshingwe Group	2.8-2.9 Ga
Mashaba and Chingezi granitic and gneissic rocks	2.9 Ga
Shabani Gneiss	3.5 Ga

2.2 GEOCHRONOLOGY

The five main tectonic units (with approximate ages, from Rb-Sr, Sm-Nd and Pb-Pb data of varying reliability) in the area are given in Table 2.2.

Table 2.3 lists the presently available geochronological information. Each group of ages represents a period of crust formation, major deformation and metamorphism. The Sm-Nd depleted mantle model ages (Chauvel et al. 1983; Taylor et al. 1991) demonstrate major periods of crustal growth at ca. 3.5 Ga, 2.9 Ga and 2.7 Ga. The age of the Mtshingwe Group is not well known, but is probably similar to the 2.9 to 2.8 Ga age of the Chingezi Gneiss terrain (see later). In contrast, the Ngezi Group is relatively precisely dated as 2692 ± 9 Ma by Pb-Pb cm-slice methods (Chauvel et al. 1983; Chapter 7). Rb-Sr ages on biotite whole-rock pairs (Hawkesworth et al. 1979) suggest that minerals over much of the southern part of the craton cooled through their blocking temperatures between 2.6-2.0 Ga ago. Biotite Rb-Sr ages in the south-east of the area (e.g. Shabani Gneiss from the Ngezi River) record dates between 2.06-1.83 Ga, synchronous with cooling events in the adjacent Limpopo Belt. Depleted-mantle model Sm-Nd ages between 2830 and 3550 Ma on argillites and cherts from the Hokonui and Manjeri Formations indicate the heterogeneity of the older crustal sources supplying sediment to the belt (Menuge 1985; Table 2.2).

2.3 THE GRANITOID CRUST

The granitoid crust exhibits a wide range of compositions and wide range of deformational states. Modal compositions of the main granitoid phases are illustrated in Figure 2.2.

2.3.1 *The Shabani Gneiss*

The 3.5 Ga gneisses east of the Belingwe Belt consists of migmatitic and isoclinally folded banded gneisses, containing abundant folded cross-cutting aplites and pegmatites as well as homogeneous foliated tonalitic or granodiorite rocks. Isoclinal folds have steeply dipping generally north-south striking axial planes. Intercalated with the gneisses are greenstone relicts, including amphibolites, ultramafic rocks, quartz schists and banded ironstones.

The outcrops in the Ngezi River have been studied in detail. Here a 200 m thick schist inclusion, intercalated within banded gneisses, contains metasediments, amphibolite and metamorphosed ultramafic rock. Quartzites contain fuchsite, and andalusite. This schist inclusion is cut by a small 3.5 ± 0.8 Ga (Rb-Sr whole-rock age) tonalite plug as described below. The schist inclusion is thought to be older than the adjacent greenstones of the Brooklands Formation, part of the Mtshingwe Group.

19

Table 2.3. Isotopic ages from the Belingwe area.

Rock unit	Method	Age ± 2σ Ma	$^{87}Sr/^{86}Sr$, ε_{Nd} or μ_1 (Pb)	Source	Comment
1. Mineral Ages	Rb-Sr mineral to w.r.				
Shabani Gneiss Zarubi Quarry	Biotite	1920		Hawkesworth et al. (1979)	
	Feldspar	1870			
	Epidote	1810			
Chibi Batholith Mukwake Quarry	Muscovite	1600		Hawkesworth et al. (1979)	Meaningless age
	Biotite	1830			
	Feldspar	3030			
Augen Gneiss Ngezi river	Biotite	2060		Hawkesworth et al. (1979)	
	Hornblende	2090			
	Feldspar	2360			
Mashaba Tonalite	Muscovite	2910		Hawkesworth et al. (1979)	Identical to whole-rock
	Biotite	2500			
	Feldspar	2450			
	Epidote	2000			
2. Great Dyke	Rb-Sr w.r.	2460 ±16	0.7026 ±4	Hamilton (1977)	
3. Chilimanzi Suite granites					
Fort Victoria area combined isochron	Rb-Sr w.r.	2570 ±25	0.704 ±1	Hickman (1978)	
Chibi Batholith Mukwake Quarry	Rb-Sr w.r.	2470 ±440	0.7047*	Hawkesworth et al. (1979)	*Initial ratio calculated at 2570 Ma
Chibi Batholith Zarubi Quarry	Rb-Sr w.r.	2560 ±220	0.7014*	Hawkesworth et al. (1979)	*Initial ratio calculated at 2570 Ma
4. Ngezi Group Greenstones					
Volcanic rocks	Pb-Pb thin	2692 ±9	8.38	Chauvel et al. (1983); Chapter 7	Model ε_{Nd} at Pb-Pb age
Belingwe	Slices	2675 ±173	8.20	Chauvel et al. (1983); Chapter 7	Samples from several greenstone
	ε_{Nd}		0.4 to 3.2	Chauvel et al. (1983); Chapter 7	belts and includes komatiites, basalts
Regional isochron	Sm-Nd w.r.	2640 ±140		Hamilton et al. (1977)	and felsic volcanic rocks
Regional isochron	Rb-Sr w.r.	2700 ±70	0.7029 ±2	Jahn & Condie (1976)	Includes few samples of Mtshingwe Group
Manjeri Formation					
Argillite	T_{DM} Sm-Nd	3320 ±100		Menuge (1985)	Crustal sources of variable age
Argillite	T_{DM} Sm-Nd	2920 ±30		Menuge (1985)	Clastic sources possibly younger
Argillite	T_{DM} Sm-Nd	2830 ±40		Menuge (1985)	than chemical sources
Ironstone	T_{DM} Sm-Nd	3410 ±70		Menuge (1985)	

Sample	Method	Age (Ma)	Initial ratio	Reference	Comments
5. Mtshingwe Group Hokonui Formation felsic volcanics	Rb-Sr w.r.	2460±600	0.706±4	Hawkesworth et al. (1979)	High initial ratio implies reset, younger than overlying strata
Argillites and cherts	T_{DM} Sm-Nd	3550±50		Menuge (1985)	Older crustal contribution to sediments, upper limit on age
	T_{DM} Sm-Nd	3090±60		Menuge (1985)	
6. Chingezi Tonalite Locality 81/8	Pb-Pb w.r.	2800±76	8.2	Taylor et al. (1991)	Combined mean Pb-Pb age of three Chingezi Tonalite suites = 2833±43 Ma
	Rb-Sr w.r.	2818±78	0.7016±2	Taylor et al. (1991)	
		3050			
Locality 81/9	Sm-Nd T_{DM}	2874±32	8.4	Taylor et al. (1991)	Three samples from tonalite clasts in Hokonui agglomerate have indistinguishable Pb isotopic compositions
	Pb-Pb w.r.	2950		Taylor et al. (1991)	
	Sm-Nd T_{DM}			Taylor et al. (1991)	
Locality 81/10	Pb-Pb w.r.	2825±100	8.1	Taylor et al. (1991)	
	Rb-Sr w.r.	2723±102	0.7015±3	Taylor et al. (1991)	
7. Chingezi Gneiss	Rb-Sr w.r.	2810±70	0.7017±6	Hawkesworth et al. (1979)	
8. Shabani-Tokwe Gneiss Complex Ngezi Tonalite Plug (see Chapter 3)	Rb-Sr w.r.	3500±800	0.701±2	Hawkesworth et al. (1979)	Larger error due to small spread in Rb/Sr ratio
Ngezi Augen Gneiss	Rb-Sr w.r.	3250±120	0.704±2	Hawkesworth et al. (1979)	High initial ratio. May be reset
Shabani area	Rb-Sr w.r.	3495±120	0.700±1	Moorbath et al. (1977)	
homogeneous and banded Gneiss	Pb-Pb w.r.	3088±46	9.0	Taylor et al. (1991)	Interpreted as reset
	Sm-Nd T_{DM}	3460		Taylor et al. (1991)	
		3460		Taylor et al. (1991)	
		3240		Taylor et al. (1991)	

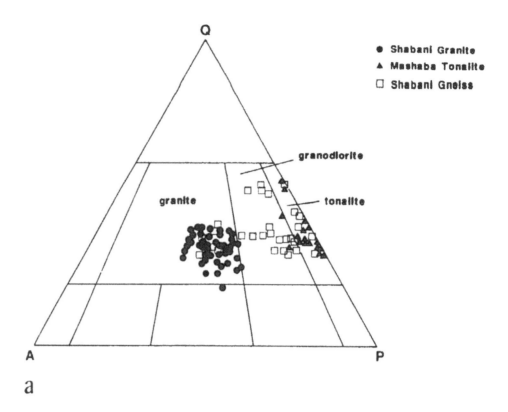

a

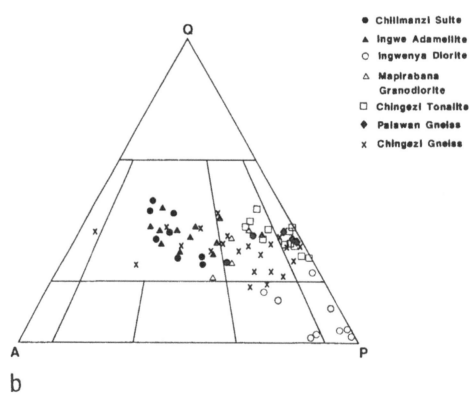

b

22

Figure 2.3. Model quartz-albite-plagioclase compositions of granitoid rocks. Fields after Streckiesen (1973). a) East of Belingwe Greenstone Belt. b) West and south of Belingwe Greenstone Belt.

Table 2.4. Regional stratigraphic nomenclature.

Fletcher & Espin (1897)	Macgregor (1947)	Stagman (1978)	Belingwe equivalents
	Shamvaian Series	Shamvaian Group	Not preserved
		Upper Bulawayan	Ngezi Group
Diorites, greenstones and ironstone	Bulawayan Series		
		Lower Bulawayan	Mtshingwe Group
	Sebakwian	Sebakwian	Shabani-Tokwe Gneiss terrain

Gneisses in the area include strongly foliated or banded tonalites, granodiorites and adamellites, cut by basic to acid dykes (Fig. 2.4). Most of the rock is banded gneiss, with 1-5 cm thick dark and light bands: Infolded into this are homogeneous grey gneisses and augen gneisses. More homogeneous granodioritic gneisses outcrop below the exposures of the basal unconformity of the Ngezi Group, along the north-west margin of the greenstone belt. These gneisses are dated at 3495 ± 120 Ma (Rb-Sr whole-rock age, Moorbath et al. 1977) and have Sm-Nd depleted-mantle model ages between 3240 and 3460 Ma (Taylor et al. 1991).

2.3.2 The 2900 Ma terrain: Chingezi Gneiss Complex and Mashaba Tonalite

2800 Ma to 2900 Ma granitoid rocks outcrop west and south-west of the greenstone belt. In the west, the migmatitic and isoclinally folded Chingezi Gneisses, with banding striking north or north-west but swinging east-west south of the greenstone belt (Fig. 2.1; Fig. 3.12; Orpen 1978), are probably a continuation of the Palawan gneiss north-west of the Great Dyke (Martin 1978, 1983). These may extend west to areas of migmatite and gneiss near Shangani mapped by Harrison (1969), as far north as the Gwenoro Dam migmatites (Stowe 1968) and the Rhodesdale Batholith, the latter dated at 2976 ± 132 Ma by the Pb-Pb whole-rock method and with a depleted-mantle Sm-Nd model age of 2990 Ma (Taylor et al. 1991). The predominantly tonalite and granodiorite gneisses exhibit a wide range in compositions (Fig. 2.2).

Chingezi Gneisses are dated at 2810 ± 70 Ma (Rb-Sr whole-rock isochron, Hawkesworth et al. 1979). The Chingezi Gneiss was intruded, prior to the Mashaba-Chibi dyke swarm, by a number of weakly foliated plutons of diorite, tonalite and adamellite which include the Chingezi Tonalite (Fig. 3.10). Sample suites from four localities of the Chingezi Tonalite have Pb-Pb whole-rock ages between 2874 ± 32 and 2686 ± 94 Ma and Rb-Sr whole-rock ages between 2818 ± 78 Ma and 2647 ± 102 (Taylor et al. 1991; all the ages are listed in Table 2.3). The most easterly locality sampled (Pb-Pb age = 2874 ± 32 Ma) intrudes the Bvute Formation and basal Hokonui Formation in addition to the Chingezi Gneiss. Large clasts of a similar tonalite are mapped within an agglomerate of the Hokonui Formation. Pb isotopic analyses of three samples of the tonalite blocks are indistinguishable from a regional Pb-Pb whole-rock isochron of all Chingezi Tonalite samples (2925 ± 30 Ma) although Taylor et al. (1991) caution against inclusion of sample suites from disparate localities in one regression. Depleted-mantle model Sm-Nd ages between 2980 and 3050 Ma confirm the inference from two-stage model μ_1 values and initial $^{87}Sr/^{86}Sr$ ratios that the '2900 Ma' event represents a major period of crustal growth. The ca. 2833 ± 43 Ma mean age of the three more easterly Chingezi Tonalite

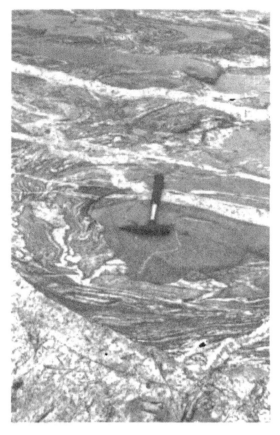

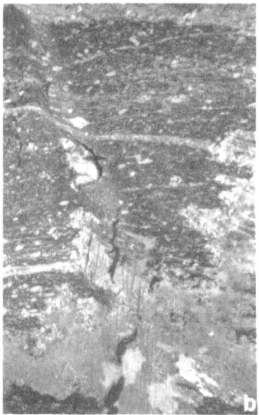

Figure 2.4. Shabani Gneiss. a) Ngezi River outcrop: Banded gneisses, intrusive phases and pegmatites. b) Augen Gneiss. Photograph about 0.5 m wide.

samples places a minimum age on the Hokonui Formation. It is possible that the Chingezi Tonalite and Hokonui Formation are cogenetic.

The Mashaba Tonalite, a composite body of massive foliated tonalite or more rarely of banded gneiss is dated at 2870 ± 160 Ma (Rb-Sr whole-rock isochron, Hawkesworth et al. 1979). It outcrops over an extensive but poorly exposed area in the east (Fig. 2.1) where it is a continuation from the type area to the east (Wilson 1968, 1973). Contact relations between the Shabani Gneiss and Mashaba Tonalite are poorly exposed although Martin (1983) describes these as gradational.

2.4 STRATIGRAPHY OF THE MTSHINGWE GROUP

The Mtshingwe Group comprises four formations: The Hokonui, Bend and Koodoovale Formations in the west, and the Brooklands Formation in the east. These formations are unconformably overlain by the Ngezi Group (Chapter 6). The nature of the basal part of the Mtshingwe Group is unclear. Orpen (1978) recognises a major unit, the Bvute Formation, of isoclinally folded and foliated amphibolite up to 5 km in width situated between the Chingezi Gneiss and the overlying Mtshingwe Group rocks. On its western contact the amphibolite is infolded with Chingezi Gneiss (Fig. 3.9). At the contact between the Bvute and the Hokonui Formations, Orpen interpreted repetitions of amphibolite mafic and felsic rock as possible interfolding between these two units. To the north Martin (1978, 1983) mapped a sequence of interbedded greenschist facies mafic and felsic rocks as alternating mafic and felsic lava flows within the Hokonui Formation, a relationship which Orpen also recognises. Towards the base of this formation, that is towards the intrusive contact of the Chingezi Tonalite, Martin records that mafic rocks predominate with metamorphic grade increasing to amphibolite facies. However in this very poorly exposed area there is no prominent foliation except in rare hornblende schists.

Two points seem clear: 1) Mafic rocks are interbedded with felsic agglomerates in the Hokonui Formation; and 2) foliated amphibolite gneisses, which are intercalated with Chingezi Gneiss on their western margin, form a major unit around the south-west margin of the greenstone belt. The possible significance of this structural setting is discussed further in Chapter 3.

The age of the Mtshingwe Group greenstones is bracketed by the intrusive and possibly cogenetic Chingezi Tonalite (ca. 2833 ± 43 Ma) and the unconformably overlying Ngezi Group greenstones dated by Pb-Pb cm-slices at 2692 ± 9 Ma (Chauvel et al. 1983; Chapter 7). The only additional age constraints on the Mtshingwe Group are the youngest detrital zircon U-Pb ion-microprobe zircon ages of 3210 ± 20 Ma from the Buchwa quartzite in the Mweza Greenstone Belt (Dodson et al. 1988). This sequence is tentatively correlated with the Brooklands Formation of the Mtschingwe Group which is in turn correlated with the Hokonui and Bend Formations (Wilson et al. 1978) and as discussed below.

2.4.1 *The Hokonui Formation*

The Hokonui Formation consists of a sequence of andesitic, dacitic and rhyolitic lavas, pyroclastic and resedimented volcanic material of similar composition, together with minor amounts of dolerites and fine-grained greenstones. The outcrop width of the formation is up to 8 km but this may include the Bvute Formation in the north and some of this thickness may reflect structural repetition. The formation is calc-alkaline, the

Figure 2.5. Tonalite blocks in Hokonui vent agglomerate, Mtshingwe River.

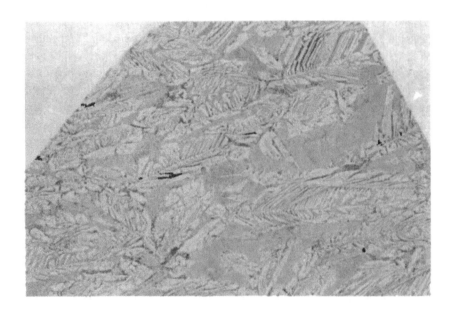

Figure 2.6. Olivine skeletal texture, komatiite, Bend Formation. Photograph is whole thin section (2.5 × 4 cm).

remains of a large volcanic pile, perhaps a single major volcanic complex.

The massive greenstones are composed of a fine-grained aggregate of tremolite-acti-nolite, chlorite, clinozoisite, albite and quartz. No relict igneous textures are preserved. Agglomerates and assorted pyroclastics and tuffs form the bulk of the formation. Clasts in these rocks range up to about 60 cm across, vary from angular to well-rounded, and are predominantly of porphyritic felsic volcanics but include occasional greenstone clasts and even fragments of bedded tuff.

In the Mtshingwe River, 4.5 km upstream of the Bulawayo-Zvishavane road bridge, is a probable vent agglomerate. This contains tonalite blocks (Fig. 2.5) up to 20 m diameter, and also smaller (1 m) blocks of felsic volcanics, set in a fine-grained recrys-tallized matrix.

In the eastern part of the Hokonui Formation, lapilli tuffs and fine-grained tuffs occur, some of the latter showing cross-bedding, load structures and fiamme structures. The source of this material was subaerial. Porphyritic felsic lavas also occur. Felsic intrusive rocks, possibly related to the Hokonui volcanicity, occur as dykes cutting the underlying Bvute Formation.

Coarse-grained massive doleritic horizons are interbedded with the Hokonui tuffs and agglomerates. No volcanic structures have been observed in these rocks, which may have been intruded as sills. Sedimentary rocks (apart from reworked volcanic material) are rare in the Hokonui Formation. They include isolated bands of arkose, a quartzite horizon, and a few thin cherts and banded ironstones. Sulphide-facies iron formation (pyrite, pyrrhotite and arsenopyrite), graphitic argillite and chert occur at the Belvedere Mine.

2.4.2 *The Bend Formation*

The Bend Formation is dominantly komatiitic. It includes repeated cycles of komatiites, komatiitic basalts and basalts, interbedded with ironstone and chert, and is capped by a major banded ironstone bed which forms Belingwe Peak. The formation overlies the Hokonui Formation in the north, and the Bvute Formation in the south. It is itself disconformably overlain by the Koodoovale Formation in the core of the Bend Syncline (Fig. 2.1). North of the Bend Syncline both the Koodoovale and Bend Formations are progressively cut out by the Manjeri Formation. The maximum present-day thickness of the Bend Formation is 6 km, in the core of the syncline, decreasing to 2-2.5 km on the limbs, but these thicknesses have been greatly affected by deformation and erosion.

Komatiites occur in stratiform bodies, probably mostly flows, ranging from a few metres to 50 m thick. Typically the basal zones of these bodies are made of pseudomorphs after cumulate olivine, while the upper parts contain distinctive pseudomorphs after coarse olivine spinifex crystals (Fig. 2.6). Some of the thicker stratiform bodies may be mapped for several km along strike. No pillowed komatiite outcrops have been identified, but in rare outcrops chert cappings exist above spinifex-textured units, suggesting that the komatiite was extrusive.

Komatiitic basalts and basalts are common, and frequently display clinopyroxene spinifex textures, with needles up to 40 cm long. Pillows are abundant, usually less than one metre across, but sometimes ranging up to 3 m in long axis. Some spherulitic pillowed basalts occur – these are typically Ti and Fe-rich compared to the majority of the Bend suite. Massive basalt and komatiitic basalt flows occur in many parts of the sequence, sometimes showing coarse-grained ophitic textures. Fresh clinopyroxenes and a fine grained groundmass after glass are preserved in the fresher samples although most mafic volcanics are metamorphosed to tremolite-epidote-chlorite-albite-sphene-quartz bearing assemblages. Olivine is invariably serpentinised. Despite metamorphic alteration igneous textures are generally well preserved.

At least ten oxide-facies banded ironstone horizons occur in the formation: These are mostly thin (2-30 m), with little lateral facies variation, but the uppermost horizon is 100 m thick. Most ironstone horizons have great lateral continuity, and the second can be traced for the entire strike length of the formation (20 km).

2.4.3 *The Koodoovale Formation*

This formation consists mainly of conglomerates, and assorted finer-grained sedimentary rocks, with a locally developed unit of felsic agglomerate. The formation occurs in the core of the Bend Syncline, above the Bend Formation, and is unconformably overlain by the Manjeri Formation which oversteps its margins. The Koodoovale Formation

27

Figure 2.7. Coarse breccia, Ndakosi Member, Brooklands Formation.

28

Figure 2.8. Deformed pillow-lava, Roselyn Member, Brooklands Formation.

ranges from 2 km thick in the core of the syncline to less than 1 km on the limbs: Much of this variation may be related to deformation and erosion. The basal contact of the Koodoovale Formation is erosive with a few major channels cut into the underlying ironstone of the Bend Formation. No angular discordance is recorded.

The conglomerate is the dominant rock type in outcrop, although this predominance may in part be a consequence of preferential exposure. The rock outcrops as lenticular bodies 200×50 m, surrounded by areas of poor exposure. Scattered outcrops in the poorly exposed areas are typically finer-grained. Clasts in the conglomerate range up to 50 cm in diameter, and are mostly moderately well-sorted and well-rounded.

Clasts were derived from most of the underlying rock types. Granitoid clasts include adamellites and tonalites (possibly including material from the Chingezi Gneiss). The common felsic volcanic clasts may come from the Hokonui Formation. Other clasts include typical Bend Formation material such as basalts, spinifex-textured komatiitic rocks, dolerites and abundant chert and banded ironstone clasts.

In the southern limb of the Peak Syncline is a locally developed felsic agglomerate, which interdigitates laterally with conglomerate. Clasts in the agglomerate are similar to Hokonui Formation felsic material.

2.4.4 *The Brooklands Formation*

The Brooklands Formation includes the supracrustal rocks lying below the basal Manjeri Formation unconformity in the southeast portion of the greenstone belt. Isolated remnants of greenstone rocks to the north may be correlated with the Brooklands Formation (Fig. 2.1). The sequence consists of a conformable set of sedimentary and volcanic rocks, thought to have been laid down on the older gneissic terrain. Detritus from this terrain is incorporated in the Brooklands breccias and conglomerates. The Brooklands Formation was deformed and eroded prior to the deposition of the Manjeri Formation, which oversteps from the gneissic basement in the north onto the Brooklands sequence in the south.

The rocks have been metamorphosed in the greenschist facies, with actinolite, chlorite and albite characteristic of meta-volcanic rocks, and phengitic muscovite and chlorite typical of the fine-grained pelites. In two localities relict andalusite and diopside occur, probably in the thermal aureole of a granite intrusion. In the following discussion non-metamorphic terminology is used.

Four members have been identified in the formation. The lowermost Ndakosi Member contains mainly sedimentary rocks, including chloritic phyllites, siltstones and conglomerates (Fig. 2.7), together with some mafic and ultramafic rocks (possibly intrusive). The Roselyn Member consists of komatiitic basalts (Fig. 2.8) and less magnesian komatiites, together with massive serpentinite bodies derived from intrusive dunite bodies of uncertain age. Above this is the Mnene River Member, which contains a distinctive and laterally continuous conglomerate (Fig. 2.9) overlain by varied fine grained and silicified sedimentary rocks, and the uppermost member is the Pemba Member, consisting of basalt and komatiitic basalt together with minor fine-grained sediment, chert and ironstone.

In the south of the area the uppermost rocks of the Pemba Member either pass without apparent discontinuity to the Manjeri Formation (although considerable erosion may have taken place) or, east of Pemba Mountain, are overlain by further ultramafic rocks, which are bounded by the Manjeri Formation.

Many later intrusions cut the Brooklands Formation: Whether parts of the Roselyn and Pemba Members are intrusive is uncertain. The correlation of the Brooklands Formation with the Bend Formation is based on structural evidence. The very considerable lateral

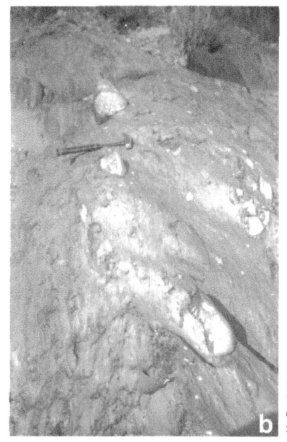

Figure 2.9. Black Prince Conglomerate, Brooklands Formation. a) General view with mafic dyke (left) cutting conglomerate. b) Detail showing rounded quartzite clast and flattened mafic clasts.

facies variation within the Brooklands Formation makes the poor lithological correlation with the distant Bend Formation not altogether surprising.

2.5 THE MTSHINGWE GROUP GREENSTONES: STRUCTURAL SETTING

2.5.1 *Western area*

In the west of the area the Bend Formation is apparently conformable upon the Hokonui Formation, although the contact is now the Mberengwa Fault and it is thus unlikely that the nature of the original contact between the two formations will ever be known. Both formations contain rocks of varied metamorphic grade (partly depending on the proximity of later granitic intrusives), with occasional relict primary minerals being preserved. The best preserved Bend Formation rocks are lowest greenschist grade komatiitic basalts. Away from the Chibi Batholith the metamorphic grade is indistinguishable from that of the Ngezi Group.

Structural work is aided by the presence of common indicators of younging direction, especially in the Bend Formation.

In the core of the syncline the Bend Formation is disconformably overlain by the Koodoovale Formation, but there is no evidence for angular unconformity or any major time-gap (see later). Both formations are folded about the Peak Syncline, plunging 65° to north-northeast. To the south the related Dube anticline folds the Bend and Koodoovale Formations.

The basal relationships between the Hokonui and Bvute Formations are not well understood as discussed above. It is also possible that the vent agglomerates of the Hokonui Formation and the Chingezi Tonalite are coeval.

2.5.2 *Eastern area*

The Brooklands Formation also displays the major syncline which is occupied by the Bend Formation, although later folding events have rotated the axis to plunge steeply north-west. The formation shows marked lateral facies variation, becoming much thicker and more varied in the core of the syncline. It is possible that the siting of the syncline may have been tectonically controlled, with the syncline forming over the site of maximum subsidence. After, or during the folding, major erosion took place; in the south the basal Ngezi Group is sub-parallel to the Brooklands strata, but to the north-east the Brooklands Formation has been entirely removed (if it were ever present), and Ngezi Group rocks lie directly on ancient Shabani Gneiss.

The correlation between the western and eastern parts of the Mtshingwe Group is on structural grounds, and the stratigraphic evidence for correlation is less good (see Chapter 3). If unfolded, Brooklands would lie 30 km or more away from the western part of the group; thus, in view of the very rapid lateral facies changes in Brooklands (see below), it is considered that despite this poor stratigraphic correlation the two rock suites may be provisionally equated, in the absence, as yet, of geochronological control.

2.6 STRATIGRAPHY OF THE NGEZI GROUP

Four formations are recognized in the Ngezi Group: The Manjeri, Reliance, Zeederbergs and Cheshire Formations. They form a relatively undeformed, coherent stratigraphic

Figure 2.10. Basal unconformity, Manjeri Formation (to right) over Shabani Gneiss (on left).

Figure 2.11. Stromatolites, Manjeri Formation, Rupemba Peak outcrop. Approx. 8 cm across block.

Figure 2.12. Reliance Formation komatiite pillow lava.

suite overlying a basement which was made of gneiss and the eroded remnants of the Mtshingwe Group. The basal contact is well exposed in at least three localities, and may be mapped and identified around most of the perimeter of the upper greenstones. The upper greenstones are dated at 2692±9 Ma (Pb-Pb isochron, Chauvel et al. 1983; Chapter 7), an age consistent with an earlier regional Sm-Nd isochron of 2640 ± 140 Ma (Hamilton et al. 1977) and regional Rb-Sr studies (Hawkesworth et al. 1975).

2.6.1 *The Manjeri Formation*

The Manjeri Formation contains the basal sedimentary rocks of the upper greenstone succession laid down on a varied terrain of tonalite, gneiss and eroded older greenstone relicts (Fig. 2.10). The formation is typically 50-100 m thick. The type section of the formation has been described by Bickle et al. (1975) and Martin (1978). At the type section basal conglomerates and quartz sandstones pass upward to varied intertidal sandstones and siltstones displaying flaser-bedding and ripple marks, for example, and associated with chert and oxide-facies banded ironstone. Above this are rocks which were probably deposited in slightly deeper water, including graded arkosic sandstone and greywacke, capped by a thin bed of sulphide-facies ironstone. The sulphide-facies ironstone is frequently sheared but the persistence of this horizon, around most of the outcrop of the Ngezi Group greenstones, suggests that no major structural break occurs at this level.

The two other localities at which the unconformity is well exposed and has been studied show a broadly similar sequence. Near Zvishavane, banded ironstone passes up into arkosic sandstone and coarse conglomerates containing clasts derived from the granite-gneiss terrain. The sequence is capped by ironstone. In the west of the belt, by the Mtshingwe River, basal conglomerates and pebbled beds pass into shallow water sandstone, siltstone and then to laminated quartzite. Sedimentary structures are common in the rocks and symmetrical ripple-mark pavements are well exposed in places. Other outcrops of the unconformity occur but have not been studied in detail.

One notable feature of the Manjeri Formation is the presence of stromatolites (Fig. 2.11) in limestone south of the type section (Bickle et al. 1975; Martin et al. 1980). These stromatolites probably formed in intertidal conditions. Much of the limestone in the formation displays possible cyanobacterial lamination, although in many areas it is too deformed to preserve depositional structures.

2.6.2 *The Reliance Formation*

The Manjeri Formation passes directly and with apparent conformity into the Reliance Formation, which consists dominantly of komatiites and komatiitic basalts (Fig. 2.12). Rocks of the Reliance Formation outcrop within less than 10 cm of the Manjeri Formation sulphide-facies ironstone at the type locality of the Manjeri Formation. Significant shear zones are present within the Reliance Formation but there is no evidence that these disrupt the distinctive stratigraphy.

The Reliance Formation contains very little intercalated sediment, apart from rare chert stringers. It is approximately 1 km thick. The most continuous exposure is at the type section, described by Bickle et al. (1975) and by Nisbet et al. (1977). The sequence is: Magnesian basalts and komatiitic basalts, passing to ultramafic pillow lavas and flows, followed by further komatiitic basalts and then tuffs and minor breccias. Several thick (up to 40 m) concordant bodies, sills or flows, occur in the succession, especially in the lower, komatiitic basalts part. These sills or flows may be traced for distances of 5 to 10 km.

Figure 2.13. Zeederbergs Formation basaltic pillow lava.

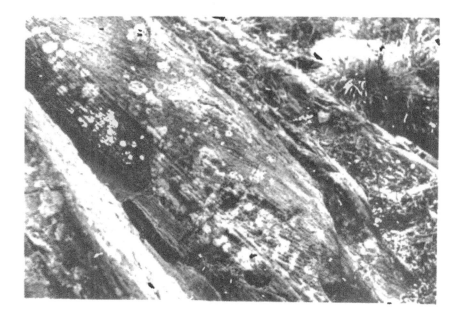

Figure 2.14. Cross-bedded sandstone, Cheshire Formation.

Clinopyroxene and olivine spinifex textures are commonly developed throughout the formation.

Metamorphism is usually in the low greenschist or sub-greenschist facies, but with considerable local variation in grade. Fresh clinopyroxene is frequently preserved. Fresh olivine is less common, but occurs in several localities, notably in the Nengerere River and near Zvishavane where the formation has been drilled (Nisbet et al. 1987). Part at least of the serpentinisation of olivine appears to have taken place during recent weathering, and fresher material may occur at depth (this recent alteration is apparent in olivine-rich rocks at the Shabani Mine to the east and the Great Dyke to the west). The petrology and geochemistry of the Reliance Formation is discussed in detail by Nisbet et al. (1977).

2.6.3 *The Zeederbergs Formation*

This formation consists of mainly basaltic lavas (Fig. 2.13) and is up to 6 km thick on both limbs of the syncline. It directly and conformably overlies the Reliance Formation.

The formation contains a typically monotonous sequence of basaltic pillow lavas, flows and possible sills, and breccias, and contains very minor intercalations of tuff and hyaloclastic debris. In the field there appears to be little significant variation in the formation from bottom to top, and available geochemical results support this. Petrographically, the extrusive lavas typically consist of needles and laths of clinopyroxene and abundant plagioclase set in a fine grained groundmass. Clinopyroxene and plagioclase microphenocrysts are common, and olivine is not found. Many rocks display coarse spherulitic textures, ocelli being slightly richer in feldspar. Spinifex textures occur, but are rare.

Very little intercalated sediment is seen in the formation, apart from rare small chert stringers, and possible sedimentary intercalations near the top of the formation. The metamorphic grade ranges up to the greenschist facies and epidote is common. Near intrusive granite hornblende-epidote-plagioclase assemblages are seen.

2.6.4 *The Cheshire Formation*

The Cheshire Formation is the uppermost formation present in the greenstone belt. It occupies the core of the syncline, for the most part resting with conformity on the Zeederbergs Formation although it contains clastic material derived in part from that formation. The maximum breadth of the Cheshire Formation is 5 km, indicating a possible thickness in excess of 2 km, although tectonic thickening and complex local structures may contribute to this.

At the base, the formation typically consists of a basal conglomerate, or limestone (in the west), or auriferous ironstone. The conglomerate contains well-rounded clasts of mafic material derived from the Zeederbergs Formation. Overlying the basal strata are argillite, siltstone and sandstone (Fig. 2.14). Locally tufts and dolerite are intercalated with the lower Cheshire Formation.

Ripple-marks are occasionally seen in the siltstone, indicating shallow-water deposition. The argillite and siltstone form much of the thickness of the Cheshire Formation. Near the top of the succession thin banded ironstones are common. The metamorphic grade is variable, but locally very low with brown kerogen present in some samples (Abell et al. 1985b).

To the west of the syncline axis is a very important limestone member containing thick algal laminated limestone. In several localities stromatolites are exposed (figured by Bickle et al. 1975 and Martin et al. 1980). Associated with the stromatolites are intercalations of sandstone and siltstone displaying common sedimentary structures including ripple marking, cross-bedding and polygonal mudcracks.

2.7 THE NGEZI GROUP: GENERAL SETTING

The Ngezi Group forms the central north-northwest trending syncline of the greenstone belt. West of the main synclinal axis, Cheshire Formation rocks are repeated by a smaller anticline and syncline. The southern closure of the main syncline can be mapped in the Cheshire Farm area, although fold closures outlined by ironstone stringers are tightly pinched. In the north the closure is obscured by deformation. As in the Mtshingwe Group, the first deformation produced little internal strain and many primary volcanic and sedimentary textures are almost perfectly preserved. Intrusion of the Chibi Batholith (Chilimanzi Suite) at 2570 Ma post-dates the synclinal folding and truncates the southern Ngezi Group.

2.8 LATE ARCHAEAN INTRUSIONS

A variety of major plutonic bodies outcrop in the area: These include both ultramafic bodies and dykes and younger granites. They have been discussed in detail by Martin (1978) and will only be briefly mentioned here.

Of the ultramafic complexes the most significant are the Shabani Complex, the Vukwe Serpentinite and the Gurumba Tumba Serpentinite (Martin 1978; Orpen 1978). Nisbet et al. (Chapter 6) present arguments to suggest that the Shabani Complex may be the remnants of a large magma chamber, in which fractionation took place from komatiitic and magnesian basalt parent liquids to erupt basalts similar to the Zeederbergs suite.

The younger granites include the Shabani Granite, part of which, known informally as Catherall's pluton, intrudes the Ngezi Group (Catherall 1973); and the Chibi Granite which has cut off the southern end of the greenstone belt, producing a marked metamorphic aureole. Both of these granites are of complex internal structure, and locally show good fabrics. Samples of this suite from south-east of the Belingwe Greenstone Belt have been dated at 2570 ± 25 Ma (Rb-Sr whole-rock, Hickman 1978).

Intrusion of the Great Dyke and the East Dyke, at 2460 ± 16 Ma (Hamilton 1977 recalculated to $\lambda_{Rb} = 1.42 \times 10^{-11}$ a^{-1}) was followed by dextral displacement on the Mtshingwe Fault and some folding in the greenstone belt. Later uplift and minor deformation probably took place in the period to 1.8 Ga, as well as various episodes of dyke intrusion.

2.9 STRATIGRAPHIC IMPLICATIONS

The discoveries in the Belingwe Belt have implications both for the tectonic significance of greenstone belts and have prompted a reassessment of the stratigraphy of the entire Zimbabwe Archaean Craton, 'casting the aeons old into another mould.'

The well preserved unconformity between the Ngezi Group greenstones and older greenstones and granitoid gneisses demonstrates that this sequence, including the komatiitic volcanics, evolved in a continental environment. The markedly different stratigraphies of the Ngezi Group greenstones and the components of the Mtshingwe Group imply that these units formed in markedly different tectonic settings: No one model is likely to account for all greenstone belts. Possible tectonic environments are discussed in the subsequent chapters.

The distinct stratigraphies allow lithostratigraphic correlations. In particular, Wilson et al. (1978), Wilson (1979) and Stagman (1978) have shown that the basic stratigraphic units of the Belingwe Belt (the lower (Mtshingwe) and upper (Ngezi) greenstones) can be traced across the craton, and marker horizons identified in most greenstone belts in these terrains.

This correlation has now been accepted by the Geological Survey as the basis for nomenclature of the Zimbabwe Archaean (Stagman 1978). It is because of the importance of the Belingwe stratigraphy that it has been described at some length above. We here propose modifications to the existing nomenclatural scheme of stratigraphy of the

Zimbabwe Archaean Craton. In particular, it has been suggested by Nisbet (1978) that the term 'Bulawayan' should be restricted to the Ngezi Group, and its equivalents, and that 'Belingwean' should be used for the Mtshingwe Group and its equivalents.

The greenstone belts of Zimbabwe were first described in the scientific literature by Fletcher & Espin (1897), who recognized a unit of 'diorites, greenstones and ironstone dykes and bars' and by Mennell (1904). By 1947 enough was known for Macgregor to set up his threefold division of Sebakwian, Bulawayan and Shamvaian Series. These were originally chronologically defined units, although they were later transformed to lithostratigraphic groups (Bliss & Stidolph 1969).

The work in the Belingwe Belt has revised this model in two ways. First, the recognition of distinct 'upper' and 'lower' greenstone units has divided what was previously thought of as the Bulawayan into two distinct and separate 'lithological' units which can be traced across the craton. Secondly, the improvement in geochronological knowledge has led to the identification of distinct chronostratigraphic units. Greenstone belts and temporally associated granitoid crust are discriminated on the basis of their age, and the basis of stratigraphic correlation is increasingly chronological, not lithological. As precise U-Pb zircon ages become available, the methods of Archaean stratigraphy must more nearly conform to the methods by which Phanerozoic correlation was first achieved in Europe and North America. These lithostratigraphic correlations within the Belingwe Greenstone Belt and with the other greenstone belts in the Zimbabwe Craton will be testable by precise geochronology.

CHAPTER 3

Structure and metamorphism of the Belingwe Greenstone Belt and adjacent granite-gneiss terrain: The tectonic evolution of an Archaean craton

M.J. BICKLE, J.L. ORPEN, E.G. NISBET & A. MARTIN

ABSTRACT

The Belingwe Greenstone Belt and surrounding terrain record a complex and diverse deformational and metamorphic history in the period from ca. 3600-2000 Ma ago. Several distinct structural units can be identified. The ancient gneissic basement to the east of the greenstone belt, the Shabani Gneiss Complex, was complexly deformed and then intruded by granodioritic bodies ca. 3500 Ma ago. About 600 Ma later, 2800-2900 Ma ago, the Chingezi Gneiss Complex formed to the west of the belt, immediately followed by intrusion of the Chingezi Tonalite to the west and the Mashaba Tonalite to the east. A complex series of deformational and metamorphic events can be recognised in the Shabani Gneiss Complex, but the intense gneiss-forming deformation which occurred in the Chingezi terrain cannot be definitely recognised in the Shabani Complex.

The lower of the two stratigraphic sequences in the Belingwe Belt, the Mtshingwe Group, may have been laid down on an older basement of the Chingezi terrain in the west and Shabani Gneiss in the east. Evidence is strong but not definite that this basal contact is an unconformity. The Mtshingwe Group was folded about a north-east striking axial plane and eroded prior to the unconformable deposition of the 2700 Ma Ngezi Group on a varied terrain of Mtshingwe Group, gneiss and tonalite. This latter unconformity is unequivocally demonstrated by good exposure. Both greenstone sequences were folded and have undergone metamorphism of varying grade, in places up to amphibolite facies but elsewhere in the zeolite facies. Intrusion of younger adamellite and dykes, including the Great Dyke, took place on the margins of the belt.

3.1 INTRODUCTION

A detailed structural understanding is a necessary prerequisite to the construction of a proper model of the geological relationships and the stratigraphy of any greenstone belt. The structural studies described below were undertaken to unravel the geological relationships of the main units. The structural work of necessity varies in detail and there is scope for further study in a number of areas. Geochronological information is necessary to complement and test geological inferences based on structural data and likewise study of structural fabrics cannot be divorced from study of metamorphic fabrics. Further work is needed in both these fields.

Table 3.1. Sequence of tectonic, depositional and intrusive events.

Time* Ma	Deformational and metamorphic events		Intrusive events	Depositional events
	D_g4	Dextral faulting and folding related to Mtshingwe Fault		
2460 ±16			Great Dyke and East Dyke	
2470 2590		Uplift and cooling from regional greenschist facies metamorphism		
	D_g3	Folding and main cleavage formation in greenstones		
2570 ±25			Chilimanzi Suite adamellite plutons	
	D_g2	North-south synclinal folding of the Ngezi Group		
2692 ±9			Mafic-ultramafic intrusion ?Mashaba-Chibi Dyke suite	Deposition of Ngezi Group
	D_g1	SW to NE synclinal folding of Mtshingwe Group		
2833 ±43			Chingezi Tonalite in west Mashaba Tonalite in east	Deposition of Mtshingwe Group
2900		Deformation and metamorphism Chingezi Gneiss terrain	Intrusion of precursors to Chingezi Gneisses	
		Amphibolite facies metamorphism		
3500 ±800		Shabani gneiss complex	Ngezi Tonalite Plug	
	D_s2	Isoclinal folding of Shabani Gneiss Complex		
c. 3500			Homogeneous Shabani plutons, aplites and pegmatites	
	D_s1	Intense flattening of Shabani Gneiss	Precursors to Shabani migmatitic gneiss	Deposition of greenstone precursors to schist inclusions in Shabani Gneiss

*Sources and details of ages tabulated in Table 2.3.

3.2 SUMMARY OF TECTONIC EVENTS AND GEOCHRONOLOGY

Our interpretation of the main tectonic events in the Belingwe area, to be justified in detail below, is listed in Table 3.1. The ca. 3500 Ma ages on banded gneisses and migmatites as well as from the more homogeneous Shabani gneisses require that the formation and major deformations to this complex took place then or earlier. The ca. 2800-2900 Ma ages on the Chingezi gneisses (in the west of the Belingwe Belt) and the Mashaba Tonalite

40

record a second major tectonic event. The complexly deformed Chingezi Gneiss suggests that a major deformational event occurred west of the Belingwe Greenstone Belt. This event has as yet not been recorded in the Shabani Gneiss to the east although the status of gneissic rocks within the Mashaba Tonalite is equivocal.

The Mtshingwe Group greenstones may have been deposited on a granite-gneiss basement comprising Chingezi Gneisses in the west and Shabani Gneiss in the east. The Mtshingwe Group greenstones have not been dated satisfactorily but our structural interpretation would place their deposition within the period 2800 to 2900 Ma. The unconformably overlying Ngezi Group greenstones, reliably dated at 2690 ± 13 Ma, were folded about a major north-northwest trending syncline (The 'Main Syncline') prior to the intrusion of the Chilimanzi Suite adamellites at about 2570 ± 25 Ma.

Rb-Sr ages on biotite whole-rock pairs (Hawkesworth et al. 1979) suggest that minerals over much of the southern part of the Zimbabwe Archaean Craton (and by inference much of the area in this study) cooled through their blocking temperatures between 2590 and 2460 Ma ago. Biotite Rb-Sr ages from the southeast of the area record ages between 2.06 and 1.83 Ga possibly reflecting events within the Limpopo Belt.

3.3 STRUCTURAL NOTATION

It is difficult to set up a consistent structural notation in a terrain in which rocks range over 10^9 years in age and which includes a wide variety of rock types. In particular, correlation from the gneisses to the greenstones is difficult. Not all areas have been studied in equal detail and the authors are aware that the deformational history of the gneissic terrain is still poorly known. We have chosen the scheme of Norris et al. (1971) for our structural notation. We recognise three structural domains: The Shabani Gneiss Complex (s), the Chingezi Gneiss Complex (c) and the greenstones (g). Deformation phases within complexes studied in sufficient detail are numbered sequentially according to terrain.

Coward et al. (1976) have attempted to correlate deformation phases across the southern part of the Zimbabwe Craton and the Limpopo Belt. They recognise four deformations; an early folding of the undifferentiated greenstone belts (F1), a regional cleavage producing deformation (F2) which strikes north-east in the east and north-west in the west, and a subsequent deformation which overprints F2 producing a crenulation cleavage (F3) and localized shearing (F4). The main penetrative cleavage-producing deformation in the Belingwe Greenstone Belt postdates two deformations in the Mtshingwe Group and one deformation in the Ngezi Group. We term the cleavage-producing deformation D_g3 (equivalent to Coward et al., F2) and the earlier deformations to the Mtshingwe Group are therefore D_g1 and D_g2. The main structural elements are illustrated in Figure 3.1 and the sequence of tectonic events in Table 3.1.

3.4 THE SHABANI GNEISS COMPLEX

Much of the ca. 3500 Ma gneiss terrain east of Belingwe consists of migmatitic isoclinally folded banded gneisses. The numerous aplites and pegmatitic cross-cutting veins are also tightly folded. Petrographically the host gneisses appear to be predominantly tonalitic or granodiorite in composition (Fig. 2.4). Isoclinal fold axial planes mostly strike north-south and dip steeply. Such migmatitic rocks are well exposed in the Runde (Lundi) River between Morton Hall Dam and the old Zvishavane-Mashava (Shabani-Mashaba) road bridge, and in the Ngezi River between the Brooklands Formation and the Chibi Ba-

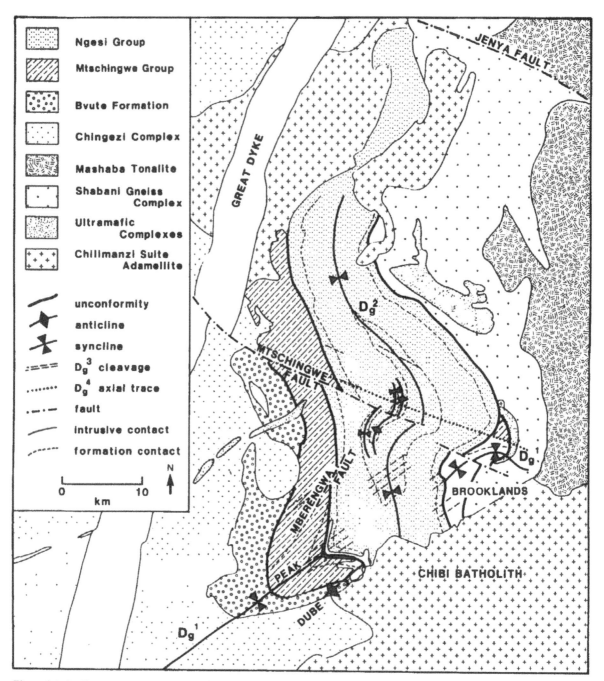

Legend:

- Ngesi Group
- Mtschingwe Group
- Bvute Formation
- Chingezi Complex
- Mashaba Tonalite
- Shabani Gneiss Complex
- Ultramafic Complexes
- Chilimanzi Suite Adamellite

- unconformity
- anticline
- syncline
- D_g^3 cleavage
- D_g^4 axial trace
- fault
- intrusive contact
- formation contact

0 — 10 km

N

JENYA FAULT

GREAT DYKE

MTSCHINGWE FAULT

D_g^2

MBERENGWA FAULT

D_g^1

BROOKLANDS

PEAK

DUBE

D_g^1

CHIBI BATHOLITH

Figure 3.1. Outline structural map of the Belingwe Belt.

tholith. Elsewhere outcrop is often poor and confined to small streams where the rocks are weathered. Infolded with the early gneisses are greenstone belt relicts which include amphibolite, ultramafic rocks, quartz schists and banded ironstone. Within the early migmatitic gneisses are more homogeneous foliated tonalite and bodies locally with minor inclusions. It is one of these that is overlain unconformably by the Manjeri Formation south of Zvishavane.

The Shabani Gneiss complex is bounded by the Belingwe Greenstone Belt to the west, the Mashaba Tonalite to the east and by younger granites of the Chilimanzi Suite to the north and south (Figs 2.1, 3.1). A well exposed section along the Ngezi River has been studied in detail and is described below. Samples of both the migmatitic Shabani Gneiss and the homogeneous rock from the area around and north of Zvishavane define an Rb-Sr whole-rock age of 3495 ± 120 Ma (Moorbath et al. 1977) with an initial $^{87}Sr/^{86}Sr$ ratio of 0.7000 ± 10. This is very similar to the age of the gneisses from the Tokwe River in the east (3500 ± 400 Ma, Rb-Sr whole-rock isochron, Hawkesworth et al. 1975 with 3560 and 3600 Ma depleted mantle model Sm-Nd ages, Taylor et al. 1991). Ages from the Ngezi outcrops discussed below are within error of these. Field distinction between ca. 3500 Ma gneisses and the ca. 2900 Ma Mashaba Tonalite is based on the transition from banded Shabani Gneiss to massive foliated medium to coarse grained Mashaba Tonalite with rare diffuse banded zones. The outcrop of Mashaba Tonalite is continuous with the type area to the east (Wilson 1973). Until the distinction between the 3500 Ma terrain and the 2900 Ma terrain is confirmed by detailed structural and geochronological studies our mapping must be regarded as a preliminary guide only.

3.4.1 *Shabani Gneisses and schist inclusion in the Ngezi River*

The section of Shabani Gneiss along the Ngezi River has been mapped in detail, both to study the structure of the gneisses and a well exposed schist inclusion and their structural relationships to the nearby Brooklands Formation (Fig. 3.2). The rock types present are gneisses, the schist inclusion and a small crosscutting tonalitic plug.

Samples of the augen gneiss described below give an Rb-Sr whole rock age of 3250 ± 120 Ma with an $^{87}Sr/^{86}Sr$ initial ratio of 0.704 ± 2. This age may be reset. The 60 by 20 m tonalite plug cutting the schist inclusion, and which post-dates two early deformations and the intrusion of aplites and pegmatites (see below) gives a poorly constrained Rb-Sr whole-rock age of 3500 ± 800 Ma with an initial ratio of 0.701.

3.4.2 *Rock types: Ngezi Schist inclusion*

The Ngezi Schist inclusion contains quartzites, psammites, pelites, banded ironstone, amphibolite and ultramafics. Bedding in metasediments is transposed into the prominent D_s1 foliation but other sedimentary structures are not preserved. Some quartzites contain fuchsite and many contain andalusite. Rare banded ironstone has a distinctive coarse quartz-magnetite mineralogy.

In the meta-igneous rocks no original textures are seen. Amphibolites range from slightly foliated and massive rocks to highly foliated rocks with alternating millimetre-thick hornblende and plagioclase rich bands. The amphibolites are cut by numerous small felsic veins. Rare thicker 10 to 30 cm tightly folded aplitic and pegmatitic veins occur. These are identical to the more numerous veins in the adjacent gneisses.

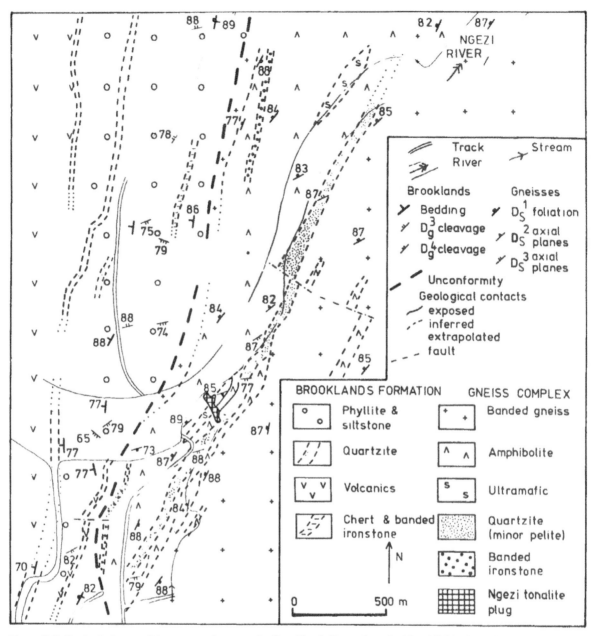

Figure 3.2. Geological map of the contacts between the Brooklands Formation, the Ngezi Schist inclusion and the Shabani Gneisses adjacent to the Ngezi River.

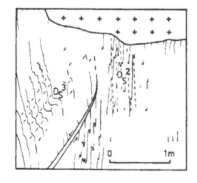

Figure 3.3. Detail of the contact between the Ngezi Tonalite Plug (ornamented with crosses) and amphibolite of the Ngezi Schist inclusion (continuous lines represent D_s1 foliation). Note D_s3 folds 'die out' towards contact with tonalite. Redrawn from field sketch.

3.4.3 Ngezi Tonalite Plug

The schist inclusion is cut by an irregular shaped intrusion (about 60 × 20 m, Fig. 3.2) of tonalite which clearly cross-cuts the foliation in the adjacent quartzites and amphibolites (Fig. 3.3) and is itself only very slightly foliated. The tonalite contains quartz-oligoclase-epidote-muscovite-biotite-chlorite and is characterized by plagioclase rich in inclusions of epidote, mica and quartz, coarse abundant epidote and partly chloritised biotite. The texture and mineralogy are very different from younger stocks of the Chilimanzi Suite.

Table 3.2. Sequence of intrusion, deformation and metamorphism in the Ngezi River section.

Deformation phase*		Deformation and metamorphism	Intrusion and deposition
(D$_g$4)	D$_s$4	Open folds with north-west striking axial planes, crenulation cleavage in Brooklands Formation	
		Greenschist facies metamorphism	
(D$_g$3)	D$_s$3	Asymmetric folds with north-east striking axial planes and axial planar shear zones in gneisses, penetrative cleavage in Brooklands Formation	
			Intrusion of Chibi batholith (contact not seen in Ngezi River – probably faulted)
D$_g$2		Synclinal folding of Ngezi Group, rotated D$_g$1 syncline in Brooklands to dip steeply and face west. Minor structures of this phase not seen in Ngezi River section, may have tightened D$_s$2 folds	
D$_g$1		Synclinal folding of Mtshingwe Group. Not seen in Ngezi River section but regional strike variation of foliation around Vukwe attributed in part to this deformation phase	
			Deposition of Brooklands Formation on gneissic basement
		Uplift and erosion	
		Amphibolite facies metamorphism	Intrusion of Ngezi Tonalite Plug
D$_s$2		Tight to isoclinal folding about upright south-west striking axial planes	
			Intrusion of pegmatites aplites (2nd set), aplites (1st set) and mafic dykes
D$_s$1		Flattening deformation and intense foliation with most earlier fabric elements destroyed	
			Intrusion of 'white-augen gneiss' into banded gneiss
		Earlier deformation – evidence rarely preserved in xenoliths in 'white-augen' gneiss & fold interference structures	Intrusion of precursors to banded gneiss, augen-gneiss and schist inclusion

*s = identified in schist inclusion or gneisses; g = in Brooklands Formation.

3.4.4 *Gneisses*

The gneisses consist of a variety of strongly foliated or banded tonalites, granodiorites and adamellites cut by a range of basic to acid dykes, the most prominent of which (forming between 10-50% of the gneisses) are two sets of aplites and a set of coarse pegmatites. Table 3.2 lists the sequence of intrusive events based on cross-cutting relationships, although only the more abundant sets of intrusives can be included in such a sequence; most compositional types apparently occur occasionally 'out of sequence.'

Banded gneiss, with 1 to 5 cm dark and light bands, comprises the bulk of the host rock. Typical minerals are quartz-albite/oligoclase-microcline-epidote-muscovite-biotite-chlorite-calcite-sphene-opaque minerals. Feldspars are rich in inclusions and the groundmass is fine-grained. Ferromagnesian minerals dominate in the dark bands.

Homogeneous grey gneisses and augen gneisses are associated with and infolded into the banded gneisses. These do not display the characteristic compositional banding although they do preserve a strong D_s1 foliation. The homogeneous grey gneiss is similar to the banded gneiss in mineralogy. The augen gneiss contains numerous 2 cm microcline phenocrysts, now augen, in a quartz-biotite-microcline-hornblende rich matrix with rarer muscovite-oligoclase-chlorite-epidote-sphene. A 'white-augen gneiss' intrudes the other gneisses. It contains conspicuous white oligoclase phenocrysts which appear less deformed than those in the augen gneiss. The matrix is fine-grained but with relatively large strained plagioclase crystals, fine quartz, oligoclase, biotite partly altered to chlorite, muscovite, epidote and rare microcline.

A small amount of amphibolite is infolded with the host gneisses. Their mineralogy is similar to early mafic dykes which cross-cut the D_s1 foliation (hornblende-plagioclase-quartz-biotite-epidote-opaque minerals).

The two cross-cutting sets of aplites are mineralogically similar, with quartz-oligoclase-microcline plus muscovite-epidote-opaque minerals ± calcite ± biotite ± chlorite. Biotite is more abundant in the younger set. Rare fine-grained porphyritic tonalite dykes contain 0.5 cm plagioclase phenocrysts set in a matrix of quartz-plagioclase-biotite-muscovite-epidote-calcite and rare K-feldspar. The abundant pegmatites cut all these dykes. They range from a few centimetres to 15 m thick. Most pegmatites appear to belong to one set though occasional cross-cutting relations are seen. They contain quartz-plagioclase-K-feldspar-muscovite. Other intrusives (possibly associated with the Chilimanzi Suite adamellites) include rare undeformed aplites, pegmatites and porphyritic tonalite. West-southwest trending amphibolite dykes of the Mashaba-Chibi suite (or Mazvihwa suite, Martin 1978) also occur in this area.

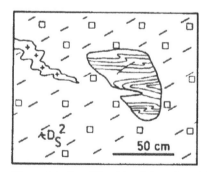

Figure 3.4. Xenolith of banded gneiss with pre-D_s2 folded foliation in white-augen-gneiss (box orna-F ment). Pegmatite (crosses) folded by D_s2. Redrawn from field sketch.

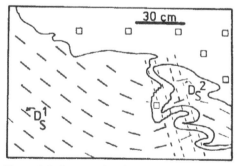

Figure 3.5. Contact of white augen-gneiss (box ornament) and banded gneiss, showing vein of white augen-gneiss folded by D_s1 and refolded by D_s2. Redrawn from field sketch.

3.4.5 *Metamorphism of the Shabani Gneisses*

Distinctive equilibrium mineral assemblages in the Ngezi Schist inclusion include horn-blende-andesine-biotite-quartz in basic rocks, quartz-andalusite-muscovite, quartz-plagioclase-muscovite-biotite and quartz-fuchsite-epidote + detrital chromite in quartzites and pelites. These assemblages indicate lower pressure amphibolite facies conditions. The amphibolite facies minerals are often partly replaced (hornblende by a tremolite/actinolite amphibole) and plagioclase is filled with fine epidote and mica inclusions. Maximum equilibrium phase assemblages in the granitoid gneisses (quartz-plagioclase-muscovite-microcline-biotite-hornblende-sphene) could also have crystallized during this early amphibolite facies metamorphic event. The inclusion filled plagioclases suggest a later greenschist facies retrogression. By contrast, little biotite shows chlorite replacement.

3.4.6 *Sequence of intrusion and deformation*

Four penetrative deformations (D_s1 to D_s4) are recognized along the Ngezi River section. In addition it will be argued that two regional folding phases (D_g1 and D_g2) must have caused major rotation of structures within the gneisses but have not caused recognizable minor structures in the rocks exposed in the Ngezi River section.

The first clearly preserved event in the Ngezi Gneisses is the intrusion of the white-augen gneiss into the banded, homogeneous and augen gneisses. Veins of white-augen gneiss are seen cross-cutting the older gneisses, and xenoliths of these are commonly preserved in the white-augen gneiss (Figs 3.4, 3.5). The xenoliths of the older gneisses exhibit an earlier folded foliation. This was followed by the D_s1 deformation which produced a fabric in both the earlier gneisses as well as the white-augen gneiss. The latter outcrops only in the eastern part of the section studied. In the western part the D_s1 deformation has rotated any earlier structures into parallelism with it and only locally can possible fold interference patterns with these earlier structure be discerned. The D_s1 foliation strikes north-east and dips steeply (Figs 3.6a, b). It is cut by the mafic dykes, aplites, pegmatites and fine-grained granites and then folded by the D_s2 deformation (Fig. 3.7).

The D_s2 folds plunge gently to steeply south-west with near vertical axial planes and a well-developed penetrative axial planar fabric (Fig. 3.6c). D_s2 folds range from isoclines in the west of the section through to more open, close, tight and occasionally isoclinal folds in the central part of the section. In the west and the central parts the D_s1 and D_s2 fabrics can only be distinguished in fold hinges but towards the east, particularly in the outcrops of the white-augen gneiss, these orientations diverge and the D_s2 foliation strikes east-northeast compared to the north-east strike of D_s1 (Fig. 3.6b). These observations suggest the strain due to D_s2 decreases to the east away from the schist inclusion and the greenstone belt. Variations in the asymmetry of the D_s2 folds suggest that the major structures are upright symmetrical folds on the scale of hundreds of metres. The amphibolite facies mineral fabrics are recrystallized subsequent to the D_s2 folding.

Superimposed on the D_s1 and the D_s2 fabrics, both in the schist inclusion and in the gneisses are asymmetric (often anticlockwise) D_s3 folds with steeply plunging axes and axial planes striking north-east (Fig. 3.6d). Axial planar shear zones are common. The D_s3 folds clearly refold D_s2 folds (Fig. 3.8) and post-date the amphibolite facies mineral fabric. The occurrence of such folds is limited to discrete zones about 1 m wide with spacing between 10 and 100 m in the western parts of the section and more in the east.

Open folds (D_s4) with east or south-east striking axial planes post-date the D_s3 folds.

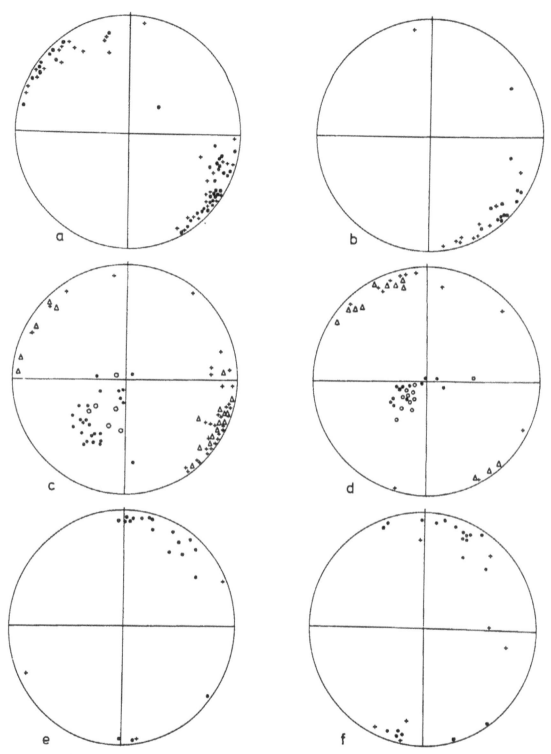

Figure 3.6. Orientation of fabric elements from the area of Shabani Gneisses and Ngezi Schist inclusion in Figure 3.13. Equal area lower hemisphere projections. a) ● Poles to D_s1 foliation in gneiss; + Poles to D_s1 foliation in schist inclusion. b) ● Poles to D_s1 foliation in gneisses west of words 'Ngezi River' on Figure 3.13; + Poles to D_s1 foliation east of words 'Ngezi River' on Figure 3.13. c) D_s2 fabric elements; ● D_s2 fold axes and + poles to axial planes in the gneisses; ○ D_s2 fold axes and △ poles to axial planes in the Ngezi Schist inclusion. d) D_s3 fabric elements; ● D_s3 fold axes and + poles to fold axial planes in the gneisses; ○ D_s3 fold axes and △ poles to fold axial planes in the Ngezi Schist inclusion. e) Poles to sinistral shears; ● in gneisses; + in Ngezi Schist inclusion. f) Poles to dextral shears; ● in gneisses; + in Ngezi Schist inclusion.

The relation between the D_s3 and the D_s4 folds and the late, undeformed rare aplites and pegmatites has not been observed. These veins, and all the earlier structures are cut by two sets of late fractures, a set striking approximately east-west (Fig. 3.6d) with sinistral displacements followed by a set striking east-southeast (Fig. 3.6e) with dextral displacements. The displacements vary from a few centimetres to several metres. The trends are close to those of shear zones mapped in the Chingezi Gneisses west of Mberengwa. The dextral shears may related to movement on the Mtshingwe Fault.

3.4.7 *The homogeneous Shabani Gneisses*

Areas of homogeneous gneiss outcrop below the Manjeri Formation south-west of Zvishavane, in the Mavinga hills east-northeast of Zvishavane and in the vicinity of the Jenya Fault (Martin 1978). In composition these range from tonalite to granite (Fig. 2.3). The mineralogy of the tonalites includes zoned plagioclase, quartz, biotite partly altered to chlorite, epidote and rare hornblende with apatite and zircon as accessories. The more granitic compositions contain large perthitic microcline and muscovite as well as plagioclase, quartz, biotite, epidote and accessories. The deformation fabric in these rocks ranges from a diffuse compositional banding, reminiscent of, but less strongly developed than that in the banded gneisses, to a simple fabric outlined by flattened quartz grains. Neither the diffuse banding nor the foliation show the tight folding characteristic of the migmatitic gneisses. The rare pegmatites and aplites which cut the homogeneous gneisses are unfolded. Contacts between the homogeneous plutons and the migmatitic gneisses are not exposed. At the unconformity localities near Zvishavane the steeply dipping foliation is truncated at the sediment contact. The lack of superposed deformations within the homogeneous gneisses would suggest that they intruded the migmatitic gneisses after D_s1 and that the formation of the foliation may correlate with D_s2 in the Ngezi River section, but this interpretation is untested.

3.4.8 *Timing of structural events in the Shabani Gneiss Complex*

The pre-3500 Ma history of the basement is not yet resolved. The 3490 ± 120 Ma Rb-Sr whole-rock age (Moorbath et al. 1977) and 3460 and 3240 Ma depleted mantle Sm-Nd model ages (Taylor et al. 1991), from a mixed suite of migmatitic and homogeneous gneiss near Zvishavane, suggests that the formation, early deformation, intrusion of aplites and pegmatites and the subsequent intrusion of the homogeneous gneisses all took place in a short time interval (less than 200 Ma) at about 3500 Ma. The ca. 3500 ± 800 Ma tonalite plug in the Ngezi post-dates D_s2 but its relationship to D_s3 is not clear. D_s3 folds 'die out' against the plug (Fig. 3.3) but the discontinuous nature of the D_s3 deformation zones may merely imply that the competence contrast between the plug and the surrounding amphibolite was important in controlling the intensity and siting of D_s3 deformation zones. Minerals in the tonalite plug appear to have recrystallized during the amphibolite facies metamorphic event characteristic of the gneisses, although partially altered to greenschist facies minerals. D_s3 folds post-date this amphibolite facies metamorphism. The intrusion of the plug is thus thought to post-date D_s2 and to pre-date D_s3.

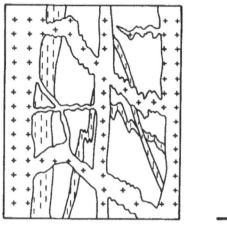

0.5 m

Figure 3.7. Cross-cutting relations between two generations of aplites (vertical dashes) and pegmatites (crosses) invading augen-gneiss (blank). Note pegmatites and aplites are folded by D_s2. Redrawn from field sketch.

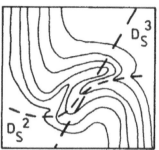

10 cm

Figure 3.8. D_s2 fold refolded by D_s3. Quartzite in Ngezi Schist inclusion. Redrawn from field sketch.

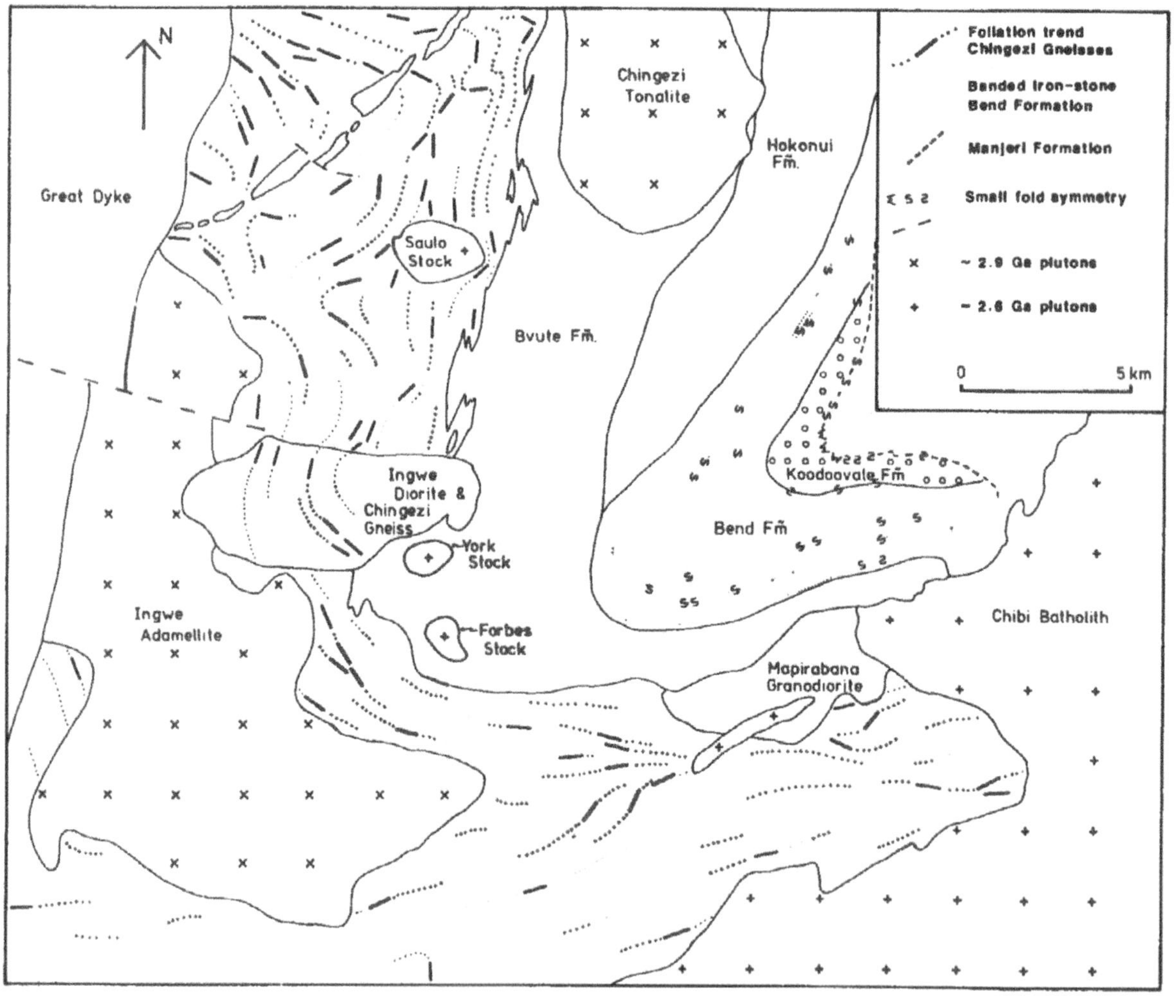

Figure 3.9. Geology of the south-west of the Belingwe Greenstone Belt. Steeply dipping foliation in the Chingezi Gneiss folded around Bend Syncline. Note small fold symmetry in Bend Formation is not related to Bend Syncline or subsequent D_g3 tightening. Mapped by J.L. Orpen.

3.5 THE 2.9 GA TERRAIN: THE MASHABA TONALITE AND THE CHINGEZI COMPLEX

Both to the west and to the east of the Belingwe Greenstone Belt geochronological work has demonstrated the existence of rocks formed at about 2900 Ma (Hawkesworth et al. 1979). Like the Shabani Gneiss Complex, rocks of ca. 2900 Ma age exhibit a range of deformational states. It appears that the more intense deformations at this time were restricted to the west of the present position of the Belingwe Greenstone Belt. More extensive descriptions of this terrain are given by Orpen (1978) and Martin (1978, 1983).

3.5.1 *The Mashaba Tonalite*

The Mashaba Tonalite outcrops over an extensive, but poorly exposed area along the eastern margin of the area studied (Fig. 3.1), and extends eastwards into the Mashava communal land where it was first recognized by Wilson (1968). The Mashaba Tonalite includes a diverse suite of medium-to coarse grained, massive, foliated and most rarely compositionally banded rocks. The mineralogy includes albite, quartz, prominent biotite and rarer muscovite, epidote and accessories.

The whole-rock Rb-Sr isochron of 2910 ± 160 Ma with an initial $^{87}Sr/^{86}Sr$ ratio of 0.7013 ± 9 (Hawkesworth et al. 1979) suggests that the bulk of the Mashaba Tonalite could not have been derived by reworking of the Shabani Gneiss terrain. It is possible that local partial melting may have taken place at the contact between the Mashaba Tonalite and the older gneiss.

3.5.2 *The Chingezi Gneiss Complex*

The Chingezi Gneiss complex comprises two main granitoid suites. The Chingezi Gneiss is a lithologically diverse group of strongly foliated, compositionally banded and isoclinally folded gneisses and migmatites, intruded by a group of slightly foliated diorite, tonalite, granodiorite and adamellite plutons and both are cut by basic dykes of the Mashaba-Chibi suite.

The Chingezi Gneiss exhibits the typical compositional and structural heterogeneity of Archaean grey gneiss units. Lenticular amphibolite bodies up to 3×0.5 m are common in some areas. A 5 to 10 cm light and dark compositional banding and parallel mineral foliation with associated leucocratic, aplitic and pegmatitic veins are all isoclinally folded on metre scales and with very variable fold axial plunges (Fig. 3.10). Interference folding is common but has not been studied in detail. The steeply dipping foliation, which is essentially axial planar to the isoclinal folding, outlines kilometre scale steeply plunging folds which mirror the main ($D_g 1$) first folding of the Mtshingwe Group greenstones (Fig. 3.9). The Palawan Gneiss which outcrops north-west of the Great Dyke exhibits a similar structural state (Martin 1983). It contains numerous inclusions of foliated amphibolite, serpentinite and iron formation.

The Chingezi Gneisses are intruded by the pre-Mazvihwa suite Ingwenya Diorite, Ingwe Adamellite, Mapirabana Granodiorite and Chingezi Tonalite. The Ingwenya Diorite intrudes the Ingwe Adamellite but the relative timing of the other plutons is unknown. The Chingezi Tonalite exhibits a very weak, steeply dipping, north-west striking foliation. On its contact with Chingezi Gneiss, Chingezi Tonalite forms a stockwork migmatite up to 5 km in width and bodies of Chingezi Tonalite truncate the isoclinally folded foliation in the gneiss (Fig. 3.11). A similar cross cutting contact is mapped between the gneiss and the foliated amphibolite of the Bvute Formation (see below). The

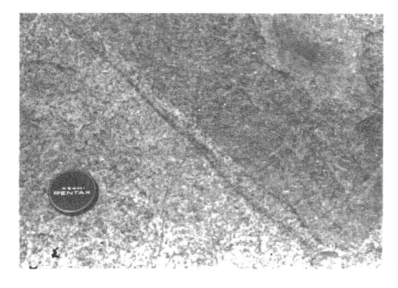

Figure 3.10. Chingezi Gneiss. GR 886284.

other plutons emplaced prior to the Mashaba-Chibi suite exhibit only sporadic weak foliations and have similar cross-cutting contacts with Chingezi Gneiss and Bvute Formation amphibolite. Several sets of late shear zones 1-2 m wide with small displacements cut all Chingezi Gneiss complex units (Fig. 3.12).

The Chingezi Gneiss has been dated at 2810 ± 70 Ma, initial $^{87}Sr/^{86}Sr = 0.7017 \pm 6$, (Rb-Sr whole-rock isochron, Hawkesworth et al. 1979) and is thought to form part of a ca. 2900 Ma terrain which continues north-west of the Great Dyke (Palawan Gneiss, Martin 1978, 1983) and north to Gwenoro Dam (Stowe 1968) as discussed in Chapter 2.

Chingezi Tonalite plutons are dated at 2800 ± 76, 2874 ± 32, 2825 ± 100, and west of the Great Dyke, at 2686 ± 94 Ma (Pb-Pb whole-rock ages, Taylor et al. 1991; see Table 2.2). These ages constrain the intense gneiss-forming deformation events west of the Belingwe Greenstone Belt to a period between 2900 and 2800 Ma. These deformation events are not recorded east of the belt.

The Bvute Formation is a major sequence of foliated amphibolite and calc-silicate (epidote-hornblende-plagioclase-calcite) rocks with an outcrop width of up to 5 km situated between the Chingezi Gneiss and Mtshingwe Group greenstones to the east. A phyllonite which outcrops west of Rupanga Ultramafic sill contains chloritoid-quartz-muscovite. Outcrop is limited to sporadic exposure in small dry watercourses. The amphibolite is isoclinally interfolded with Chingezi Gneiss on the western contact and is cut by folded pegmatite and aplite veins (Fig. 3.13). Sporadic layers of Chingezi Gneiss outcrop within the Bvute Formation. As in the underlying Chingezi Gneiss the isoclinally folded foliation in the Bvute Formation is folded around the Bend Syncline with occasional development of a crenulation cleavage axial planar to the syncline (Fig. 19 in Orpen 1978; Fig. 3.14). The Bvute Formation is intruded by Chingezi Tonalite and develops an unfoliated hornfels amphibolite on the contact. The contact between underlying Hokonui and Bend Formations is very poorly exposed and contact relations are discussed further below.

3.6 THE MTSHINGWE GROUP

The Mtshingwe Group contains the Hokonui, Bend and Koodoovale Formations in the west, and the Brooklands Formation in the east.

The Bend Formation is preserved in the Peak syncline (Figs 3.1, 3.9) (Orpen 1978). Numerous indicators of younging direction confirm that the syncline faces northeast. In the core of the syncline the Bend Formation is overlain disconformably by the Koodoovale Formation. Erosion at the base of the Koodoovale Formation occurred, but there is no evidence for a major angular unconformity (Chapters 2, 4).

The Hokonui, Bend and Koodoovale Formations are overlain unconformably by the Manjeri Formation of the Ngezi Group Greenstones (see Chapter 6). The major syncline (the Peak Syncline) plunges about $65°$ to the north-northeast (Fig. 3.15). South of this a complementary anticline folds the Bend and Koodoovale Formations. This deformation (D_g1) must have been accompanied by, or followed by, erosion of much of the Koodoovale and Bend Formations and part of the Hokonui Formation as the Manjeri Formation rests unconformably on all of these formations. The D_g1 folding, in common with early deformations to many greenstone belt sequences, has not imparted any deformation fabric to many of the rocks.

In the west, the second main deformation to the Mtshingwe Group is about the north-northwest D_g2 synclinal folding of the Ngezi Group greenstones. This has folded the D_g1 syncline into a type 1 interference pattern (Ramsay 1967) with the complementary

Figure 3.11. Chingezi Tonalite intruding Chingezi Gneiss. Tonalite intrudes parallel to the foliation at right edge of photograph and cuts directly across foliation. GR 886284.

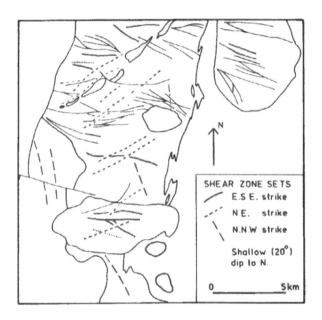

SHEAR ZONE SETS

E.S E. strike

N E.   strike

N.N.W strike

Shallow (20°)
dip to N.

0   5km

Figure 3.12. Shear zone trends in Chingezi Gneiss Complex.

south-easterly facing portion preserved in the Brooklands Formation to the east, described below (Fig. 3.1). The final D_g3 penetrative deformation tightened the Bend syncline, folded the overlying Manjeri Formation and imparted a north-east striking cleavage to the Hokonui and Koodoovale Formations (Fig. 3.16).

Rocks of the Mtshingwe Group have been metamorphosed to greenschist facies assemblages although relict igneous textures and minerals are commonly preserved including a plagioclase with An_{62} in a sample from the Bend Formation (Orpen 1978). In the Hokonui Formation relict igneous clinopyroxene and andesine are observed. In the Bend Formation too, relict igneous clinopyroxene is common in the komatiitic lavas. The maximum equilibrium metamorphic phase assemblage recorded in most of the basic rocks is actinolite-chlorite-albite-epidote-quartz-calcite-sphene (Orpen 1978). It is not possible to determine if the rocks of the Mtshingwe Group were metamorphosed prior to

54

the deposition of the Ngezi Group greenstones since, away from granite aureoles their present metamorphic grade is indistinguishable from that of the Ngezi Group.

3.6.1 *Structural setting of the Western Mtshingwe Group*

The internal structure of the Mtshingwe Group and its structural relationship to the Chingezi Gneiss complex and Bvute Formation present a number of unresolved problems in structural interpretation.

The Peak Syncline and associated anticline (D_g1) constitute the most clearly recognised deformation in the Mtshingwe Group. However the Bend Formation contains a few well foliated talc- tremolite-chlorite schist horizons (especially well developed below the second banded ironstone horizon) in which the foliation appears to predate and be folded by the D_g1 Peak Syncline (Fig. 3.16b). Furthermore the symmetry of small folds in Bend Formation banded ironstone horizons does not reflect the expected symmetry variations of the D_g1 folding although it is difficult to determine the relative timing of folds in such cherty rocks. It is possible that significant pre-D_g1 shearing took place in the Bend Formation. The Koodoovale Formation exhibits a well-developed axial plane cleavage fan in pelitic rocks (Fig. 3.16c). This cleavage may relate to D_g1 or more likely to the superimposed and co-axial/coplanar D_g3 folding which deforms the unconformably overlying Manjeri and Reliance Formations and must have tightened D_g1 folds.

The contact between the Hokonui and Bend Formations is not well exposed and may be sheared. The internal structure of the Hokonui Formation is also somewhat uncertain. Martin (1983) mapped alternating mafic and felsic volcanic rocks (Chapter 4) in the north as an intercalated volcanic sequence which faces uniformly eastwards. In contrast, Orpen (1978) interpreted the alternating amphibolitic and felsic volcanic rocks mapped by Martin (1978) as an infolded sequence. Exposure is poor. Orpen correlated some of the mafic rocks as infolded antiforms of the underlying Bvute Formation. The Bvute Formation can be distinguished from low grade mafic volcanic rocks of the Hokonui Formation by its deformational state and higher (amphibolite facies) metamorphic grade. It is not certain if this difference reflects a difference in age. Dykes of felsic igneous material, possibly feeders to the Hokonui extrusives, are mapped cutting foliated Bvute Formation amphibolites.

This problem of the basal relationship of the Western Mtshingwe Group is typical of many (if not most) greenstone belts. The rapid transition from low grade, low strain greenstones to high-grade high-strain greenstones and gneisses might represent tectonic strain and metamorphic grade variations imposed on an initially intrusive contact between granitoid and greenstone. Such a relationship has been demonstrated in the 3450 Ma greenstones in the Pilbara, Western Australia (Bickle et al. 1980) and there it has proven difficult to identify equivalent deformation events in high and low grade areas although stratigraphic greenstone units can be traced from low to high strain areas (Bickle et al. 1985).

An alternative possibility is that the Hokonui Formation was deposited unconformably on an already highly deformed basement of Chingezi Gneiss and Bvute Formation basement. The felsic dykes in the Bvute Formation suggest the second interpretation is correct although this is untested by geochronology or geochemistry.

The Chingezi Tonalite intrudes the Bvute Formation and the basal Hokonui Formation. In the latter it creates a very poorly exposed hornfelsic aureole of feldspathic amphibolite. The Pb-Pb whole-rock ages on the Chingezi Tonalite (mean 2833±43 Ma), define a minimum age for the Hokonui Formation. However blocks of tonalite in a Hokonui agglomerate are petrologically and isotopically similar to the Chingezi Tonalite (Chap-

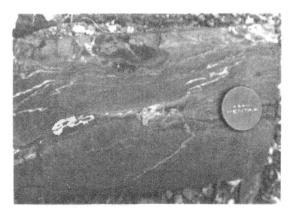

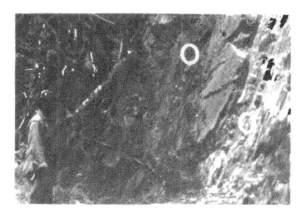

Figure 3.13a. Tightly folded quartzo-feldspathic veins in amphibolite.

Figure 3.13b. Contact of Chingezi Gneiss and Bvute Formation GR 943222.

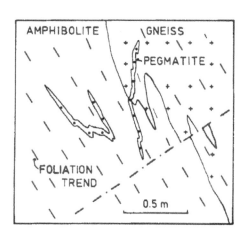

Figure 3.13c. Contact between the Bvute Formation amphibolite (left) and Chingezi Gneiss (right). Note cross-cutting pegmatite veins folded with foliation in amphibolite and gneiss axial planar to folds. Redrawn from field sketch.

ters 2, 4) which therefore may be a granitoid stock cogenetic with the volcanic edifice. The interpretation of an unconformable contact of Hokonui Formation and the Chingezi Gneiss would constrain the Hokonui Formation to be younger than the rather poorly defined ages of the Chingezi Gneiss (2810±60 Ma, Rb-Sr whole-rock).

3.6.2 *The Brooklands Formation*

The Brooklands Formation is best preserved in the south-east of the Belingwe Greenstone Belt where it comprises a steeply dipping sequence of mainly clastic sediments, ironstones and komatiitic volcanics. Coarse sedimentary breccias contain rare clasts of banded gneiss which may possibly be derived from the 3500 Ma gneiss terrain (Fig. 5.3). The Brooklands Formation has a complex stratigraphy, the most marked feature of which is the very rapid lateral thickness changes in some units. Way-up structures such as pillowed volcanics and cross-bedded sandstones all young to the west. The main structural features are shown in Figure 3.17.

56

3.6.3 *Structure and metamorphism of the Brooklands Formation*

The oldest structure preserved in the Brooklands Formation is the steeply plunging D_g1

syncline also seen in the Bend Formation to the south-west of Belingwe. The Brooklands Formation shows a marked increase in thickness and sedimentary complexity towards the core of this syncline which controlled sedimentation. Erosion must have accompanied or followed this folding. In the south the basal sediments of the Ngezi Group greenstones strike sub-parallel to the strata in the underlying Brooklands Formation. To the north the Manjeri Formation cuts progressively lower into the Brooklands Formation and north of the D_g1 syncline axis sedimentary beds in the Brooklands Formation strike at a high angle to the basal Manjeri Formation of the Ngezi Group. North of Vukwe Mountain the Brooklands Formation has been eroded completely (except for a few areas of outcrop of uncertain correlation) and the Manjeri Formation rests directly on the 3500 Ma gneisses.

The main synclinal folding of the Ngezi Group (D_g2) rotated the D_g1 syncline in Brooklands to plunge vertically and face south-east. This deformation was followed by intrusion of the Chibi Batholith in the south and of a similar stock of younger granite north of Vukwe Mountain. The Chibi Batholith truncates the Brooklands Formation in the south. A metamorphic aureole was developed adjacent to the Chibi Batholith but minerals in this aureole (probably cordierite in pelites) have been pseudomorphed during a subsequent greenschist facies metamorphism. Relicts from an early metamorphism are found in two other areas of the Brooklands Formation. Diopside in a calcareous quartzite, and andalusite (mostly or completely pseudomorphed by muscovite) in pelites, are found in the areas close to the Ngezi River just above the basal contact with the 3500 Ma Ngezi Schist inclusion. Pseudomorphs filled by muscovite, similar to those after andalusite, are also found in outcrops of pelite in the Ngezi River just above the outcrop of the East Dyke in the Ngezi River (but well outside any likely metamorphic aureole of the dyke). It is possible that this early metamorphism is related to shallow stocks of younger granite, the sporadic and localised occurrence of such early minerals being indicative of a contact-related rather than regional metamorphism.

The D_g3 deformation strikes south west. The associated cleavage is distorted around these early metamorphic porphyroclasts (including those in the metamorphic aureole of the Chibi Batholith). This deformation formed the prominent south-east striking penetrative cleavage found over most of the Brooklands Formation and must have tightened the D_g1 syncline. D_g3 also formed larger scale folds with wavelengths of up to 1 km in the overlying Manjeri Formation. These folds are best preserved around Pemba Mountain where ironstones also show numerous small folds attributed to this generation as well as some earlier tight folds of uncertain age. The main greenschist facies metamorphism (quartz-muscovite-chlorite-albite in pelites, chlorite-actinolite-epidote-albite in mafic rocks) took place after D_g3 with muscovite in particular growing on the well defined D_g3 cleavage. In the north of Brooklands in the area of the D_g1 syncline the north-west trending D_g4 folding has rotated the D_g1 syncline to face west and also the D_g3 cleavage to strike west or even west-north-west (Fig. 3.17). The D_g4 deformation has deformed the greenschist facies mineral fabric, producing a prominent crenulation cleavage in micaceous lithologies. The north-west trending Mtshingwe Fault has caused about 1 km dextral displacement across the centre of the Brooklands outcrop. The combined displacement of D_g4 and the Mtshingwe Fault displaces the Great Dyke about 4 km with no evidence of D_g4 folding. It is possible that brittle faulting of the Great Dyke and the granite-gneiss terrain to the west was taken up by more plastic D_g4 folding in the greenstone terrain. D_g4 and faulting on the Mtshingwe Fault may have been synchronous, part of the same event.

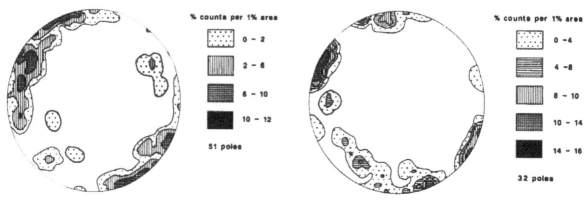

% counts per 1% area

0 - 2
2 - 6
6 - 10
10 - 12

51 poles

% counts per 1% area

0 - 4
4 - 8
8 - 10
10 - 14
14 - 16

32 poles

Figure 3.14. Poles to foliation, Bvute Formation.

Figure 3.15. Poles to bedding, Bend Formation.

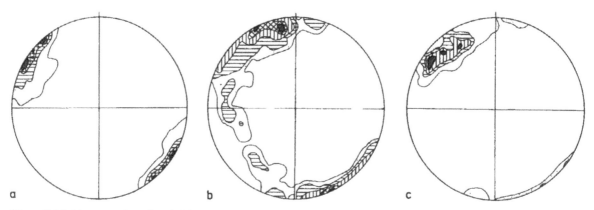

a

b

c

Figure 3.16. Cleavage orientations in Mtshingwe Group greenstones from the southwest of the Belingwe Greenstone Belt. a) Hokonui Formation (contours 8-16-24-28%). b) Cleavage in talc-tremolite schists, Bend Formation (contours 2-6-10-16%). c) Koodoovale Formation (contours 8-12-16-24%). Note: Contour values % per 1% area.

3.6.4 *The basal relations of the Brooklands Formation*

The Brooklands Formation abuts 3500 Ma Shabani Gneisses in the southern part of its outcrop area. Further north it is juxtaposed against amphibolites and metasediments of the schist belt inclusion (Fig. 3.2). North of the Mtshingwe Fault the Brooklands Formation overlies the massive serpentinite of Vukwe Mountain, except in one poorly exposed locality where banded gneisses are exposed in an old mine shaft just west of Vukwe serpentinites. The Vukwe Ultramafic is considered to be related to the Mashaba Ultramafic suite. The contact between the sediments of the Brooklands Formation and the gneisses or the schist belt inclusion is not exposed. North-south trending faults in the Chibi Batholith to the south are visible on air photos and it is likely the contact is displaced by later faulting. Faulting on several parallel faults in the north appear to be related to the Sabi Shear zone (Fig. 2.1) which causes 400 m sinistral displacement of the Manjeri Formation.

The best exposures across the contact of the Brooklands Formation and 3500 Ma basement are in the Ngezi river. Here exposures of relatively undeformed banded psammites outcrop within 10 m of isoclinally folded foliated amphibolites. From our correlation of deformation fabrics, the earlier deformations in the schist inclusion rocks are not apparent in the Brooklands Formation sediments. The first penetrative deformations in

58

the Brooklands Formation, the D_g3 cleavage and the D_g4 crenulation cleavage events, are correlated in the schist inclusion with the post-amphibolite-fabric D_s3 and D_s4 deformations, which are associated with retrograde greenschist facies assemblages. The D_g1 synclinal folding in the Brooklands Formation is reflected by a sympathetic change in strike of the underlying gneisses south and east of Vukwe Mountain. The tight D_s1 and D_s2 folding events and the aplites and pegmatites in the schist inclusion and gneisses were rotated by D_g1. D_g2, the first synclinal folding in the Ngezi Group greenstones, caused the D_g1 syncline in the Brooklands Formation to plunge steeply and face west and probably caused a similar rotation in the underlying 3500 Ma gneiss complex, but with no identifiable smaller scale structures.

Thus, on structural grounds, we argue that the gneiss complex exhibits both earlier deformations and an earlier higher grade metamorphism than are seen in the Brooklands Formation. The presence of the few early relict amphibolite facies minerals in Brooklands complicates arguments relating to the structural correlation between Brooklands and the gneisses. It is possible that the Brooklands Formation has been downfaulted against the gneisses and that the schist belt inclusion represents sedimentary and volcanic Brooklands rocks strongly deformed and metamorphosed at depth. The sedimentology, and correlation of the D_g1 syncline with that in the Mtshingwe Group in the west of the Belingwe Greenstone Belt, provide additional evidence that the Brooklands Formation is younger than the 3500 Ma gneiss complex.

The sediments in the basal member of Brooklands, conglomerates, breccias and particularly arkose and quartzites are consistent with shallow water sedimentation adjacent to a continental landmass (Chapter 5) and the banded gneiss clast in a breccia provides further support for this hypothesis (Fig. 5.3). Furthermore, if the D_g1 synclinal folding of Brooklands correlates with the D_g1 synclinal folding of Mtshingwe Group greenstones in the west and these were formed between 2900 and 2800 Ma as discussed above, then the first deformation to the Brooklands Formation was younger than ca. 2900 Ma.

3.6.5 *Correlation of the Mtshingwe Group greenstones*

The Brooklands Formation and the Bend and Koodoovale Formations can be correlated on structural grounds, the first deformation to all three being the D_g1 folding. Stratigraphically the correlation is less good. Only one outcrop of pyroclastic or felsic volcanics possibly equivalent to those in the Hokonui Formation has been mapped in the east of Belingwe. However, the Hokonui Formation, as is typical with felsic volcanoes (eg. Barley et al. 1979), thins significantly southwards on the western margin. Before the D_g2 folding Brooklands might have been more than 30 km from the western outcrops of the Mtshingwe Group greenstones. The stratigraphic correspondence between the Bend and Brooklands Formation is limited. All that can be said is that, in both, komatiitic volcanics are present, intercalated with banded ironstones. The basal largely sedimentary member in Brooklands with its distinctive quartzites and conglomerates is absent in the Bend Formation. However these differences may reflect a transition from the margin to the centre of a rifted depositional environment. The Mweza Greenstone Belt southeast of the Belingwe Greenstone Belt contains a succession passing from quartzites up into alternations of meta-volcanics and banded ironstones similar to the Bend Formation (Worst 1956). Since the Bend and Brooklands Formations may be correlated on structural grounds, and given the rapid lateral facies variation of the Brooklands Formation, we consider it possible that the growth of the Hokonui volcano, followed by the deposition

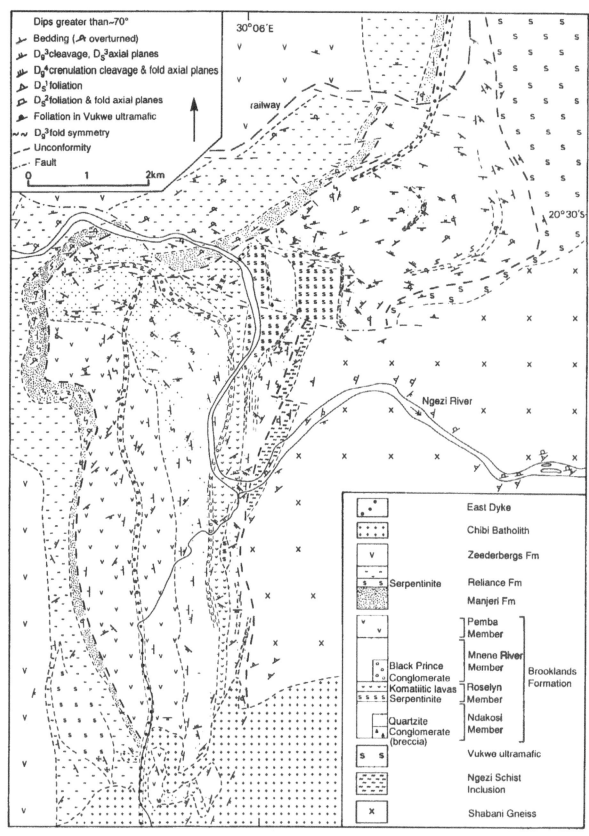

Figure 3.17. Map of structural elements in the Ngezi Group, Brooklands Formation and Shabani Gneiss Complex south-east of the Belingwe Greenstone Belt.

of the Bend and Brooklands Formations may have taken place in a relatively short interval in the same evolving tectonic setting.

3.7 THE NGEZI GROUP

Greenstones of the Ngezi Group, are preserved in the central north-northwest trending syncline. At the base the thin (ca. 150 m) shallow water sediments of the Manjeri Formation have been mapped around much of the greenstone belt, as have the 1 km thick komatiitic basalts and komatiites of the Reliance Formation, and the 6 km of largely tholeiitic pillowed basalts (the Zeederbergs Formation). The centre of the syncline is occupied by the sediments of the Cheshire Formation, which include conglomerates derived from the underlying volcanics as well as siltstones, stromatolitic limestones, banded ironstones and quartzites.

3.7.1 *Deformation history of the Ngezi Group*

The first deformation to the Upper Greenstones was the formation of the north-northwest trending D_g2 syncline. West of the main syncline axis the sediments of the Cheshire Formation (with thick stromatolitic limestones) are repeated by a smaller D_g2 anticline and syncline (Fig. 3.1). The southern closure of the Manjeri and Reliance Formations is cut out by the Chibi Batholith which thus postdates D_g2. This southern closure is well preserved within the Cheshire sediments where distinctive sedimentary lithologies including a banded ironstone and an arkose can be mapped around the syncline axis which here plunges steeply north. In the north of Cheshire the rocks have been highly sheared. They are more deeply eroded and the northern end of the exposure represents a lower structural level. North of this the Belingwe Greenstone Belt has been intruded by younger granites and the rocks are much more deformed and metamorphosed to amphibolite facies rather than the greenschist facies characteristic of the rest of the greenstone belt.

The Chibi Batholith, a member of the Chilimanzi suite (Wilson 1979) cuts the Belingwe Greenstone Belt in the south. The granite predates D_g3. The Shabani Granite, a correlative of the 2570± 25 Ma Chilimanzi Suite, outcrops extensively around the northern more highly deformed part of the greenstone belt. The granite, which is little deformed, clearly intrudes and crosscuts parts of the more foliated greenstones. The high strain event in the greenstones must therefore predate granite intrusion. The northern fragment of the belt is the only area where D_g2 fabrics are widely developed and Martin (1983) suggests that this area of the greenstone belt represents a deeper erosional level, where the basal 'pinched' closure to the syncline is exposed (Fig. 3.18). Downfolding to 10 km depth could generate amphibolite facies metamorphic assemblages.

The first deformation has not resulted in a significant internal strain to the rocks, except in the northern part of the greenstone belt. Igneous textures and volcanic and sedimentary structures are often preserved undeformed. The strain associated with this first phase of isoclinal folding may have been taken up along the narrow shear zones. These have been mapped at various locations in the volcanics. It is likely that these shear zones were reactivated during the subsequent deformations and estimates of displacement sense on these shear zones have not proved possible. Small folds related to D_g2 have only been found at three localities, all within the Cheshire Formation. At two of these localities, in the Ngezi River on the east limb of the syncline and at the southern end of the limestone horizon on the western limb, these fold axes are sub-horizontal, although they show local variations in plunge of up to 30° about the horizontal. The third locality is south of the

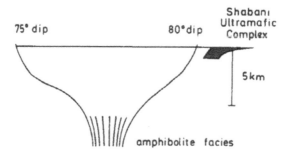

Figure 3.18. 'Champagne Glass Model.' Schematic section through Belingwe Greenstone Belt based on assumption that northern part of greenstone belt was buried ca. 10 km deeper than lower grade central part, on dips as observed at the surface and gravity data (Podmore et al. 1984; Podmore, in prep.) which indicates a maximum average depth of ca. 5 km for the greenstones. Dip of Shabani Complex after Laubscher (1963) possibly indicates rapid shallowing of palaeo-horizontal away from greenstone belt and may indicate steep dips along present margin are a consequence of subsequent deformations.

main closure of the banded ironstones on Cheshire ranch and here the D_g2 fold axes plunge steeply north. The small anticline and syncline west of the main syncline axis also plunge shallowly. This would suggest that the D_g2 plunge is nearly horizontal over much of the central part of the belt and only steepens rapidly near to the southern closure. It is not possible to determine whether the plunge variation is an original D_g2 feature or has been subsequently superimposed on the D_g2 syncline. In particular it is difficult to assess the deformation associated with the Chibi Batholith to the south. Present contacts with the Chibi Batholith are mostly sharp and cross-cutting but in the southern part of the Brooklands Formation small stocks of younger granite separate narrow, more highly deformed zones of the Brooklands Formation. Much of this deformation is related to the D_g3 cleavage-forming event.

3.7.2 Metamorphism of the Ngezi Group

The timing and distribution of metamorphic events within the Ngezi Group is problematic for a number of reasons: 1) The monotonous lavas of the Zeederbergs Formation have not been mapped in the same detail as other areas; 2) it is probable that the Ngezi Group underwent up to four distinct metamorphic events excluding modern weathering; and 3) the massive mafic lavas do not contain deformational fabrics which can be used as time markers to separate different metamorphic events.

The Ngezi Group lavas probably underwent localised but intense early hydrothermal alteration on eruption in addition to a pre-Chilimanzi Suite upper-greenschist to amphibolite facies metamorphism in the northern part of the belt (the Snake's Head), a localised amphibolite facies thermal aureole adjacent to the pre-D_g3 Chilimanzi Suite granites both in the north and south of the belt, and a post D_g3 regional event of greenschist facies (or sub-greenschist facies in parts of the Cheshire Formation) over the whole belt.

In the north of the belt foliated upper greenschist (actinolite-epidote-quartz-sphene) and amphibolite facies (hornblende-partly altered andesine) rocks are cross-cut by the Chilimanzi Suite intrusives. The foliation and associated metamorphism are thought to have formed as a consequence of the D_g2 folding of the belt.

A 1 to 2 km wide upper greenschist to amphibolite facies (hornblende-plagioclase-quartz-chlorite-sphene-opaques) thermal aureole is developed adjacent to the Chibi Batholith along the southern margin of the belt. The presence of numerous granite stocks within this aureole suggests that the granite contact is shallowly dipping and the thickness of the aureole may be much less than its width on the ground.

The rest of the greenstone belt is characterized by greenschist or sub-greenschist facies assemblages. Pelitic rocks in both the Manjeri and the Cheshire Formation contain very fine grained chlorite-sericite-quartz assemblages. The Reliance and Zeederbergs Formations contain widespread relict igneous assemblages variably replaced by greenschist facies mineral assemblages. The least altered volcanic rocks are komatiitic lavas from near Zvishavane (Nisbet et al. 1987) which are as fresh as the Mesozoic Gorgona Island komatiites with olivine very little altered, fresh augite and magnesian pigeonite and glass with low water contents (see Chapter 8).

Most volcanic rocks exhibit rather more extensive alteration in which olivine has been substantially to completely replaced by serpentine-magnetite or chlorite, clinopyroxene partly altered to actinolite, bastite pseudomorphs formed after pigeonite or orthopyroxene and plagioclase altered to albite. Doleritic and basaltic rocks are variably altered to actinolite-chlorite-clinozoisite or epidote-albite-sphene assemblages. Partially altered olivine kernels are common in the thicker cumulate portions of komatiite flows and relict olivine is also found in four spinifex komatiite localities. Unaltered clinopyroxene is widespread and unaltered igneous plagioclase microlites are preserved in a few basalts. The greenschist facies alteration could result from both syn-depositional hydrothermal alteration (by analogy with modern lava sequences) or from the post-D_g3 regional greenschist facies metamorphism demonstrated from textural evidence in the Brooklands Formation (Mtshingwe Group). Possible detrital epidote, in Cheshire Formation conglomerates (Martin 1978) provides evidence for early alteration.

Carbonate rocks from the stromatolite locality in the Cheshire Formation provide evidence of very low metamorphic grade in local regions of the central part of the belt. Some of the carbonate rocks are essentially unrecrystallised and preserve the micritic carbonate textures. The presence of dark brown kerogen, of ^9C-^{20}C aliphatic hydrocarbons and the H/C ratio of 0.1 in the kerogen (Abell, pers. comm.; Hayes et al. 1983) would suggest that metamorphic temperatures at this locality did not exceed 150-200°C or less (Abell et al. 1985b). This conclusion is supported by the results of stepped temperature pyrolitic release of nitrogen from the kerogen which has shown (Gilmour, pers. comm. 1985) that modern atmospheric nitrogen is released below 200°C but ancient nitrogen is lost rapidly at 300–400°C. In contrast kerogen from the Manjeri Formation stromatolite locality is black. Doleritic rocks in the Cheshire Formation are variably carbonated and contain turbid plagioclase, uralitized pyroxene, quartz and ilmenite altering to leucoxene. The lack of epidote is consistent with sub-greenschist facies conditions.

3.7.3 *Basal relations of the Ngezi Group*

The basal sedimentary Manjeri Formation of the Ngezi Group rests unconformably on the 3.5 Ga gneisses to the east, oversteps onto the Brooklands Formation in the south-east and rests on the Hokonui and Bend Formations in the west. The actual contact is well exposed in four localities, at two of which the Manjeri Formation lies on Shabani Gneiss, one on Hokonui Formation and one above the Bend Formation (Chapter 6). A marked discordance is mapped between the Brooklands Formation and the overlying Manjeri Formation and there is every reason to suppose that the Ngezi Group strata were deposited unconformably on this formation.

The presence of komatiite lavas in the Reliance Formation overlying the Manjeri Formation thus implies that these lavas were extruded above continental crust. Since this is one of very few localities where such a relation has been established, the possibility of tectonic juxtaposition must be considered carefully. Burke et al. (1976), for example, point out that in many orogenic belts narrow shear zones (less than 1 m thick) may separate thick thrust slices and that similar shear zones may exist but be hidden by the poor exposure in greenstone belts. Although continuous exposure exists from the unconformity across most of the Manjeri Formation, to the contact between it and the Reliance Formation, occasional stratigraphic breaks are present in the upper part of the Manjeri Formation. In several localities, notably in the type section of the Manjeri Formation, and in the type section of the Reliance Formation on the Lou Estate, undeformed mafic volcanics outcrop within a few centimetres to a metre of the top gossan-like sulphide-bearing banded ironstone of the Manjeri Formation. The great continuity of stratigraphy within the Reliance, Zeederbergs and Cheshire Formations (which vary markedly in competence) argues against tectonic emplacement of these units. The Reliance Formation, mostly a 1-1.5 km thick sequence of distinctive komatiites and komatiitic basalts, generally overlies the Manjeri Formation and underlies the Zeederbergs Formation. It exhibits a distinctive stratigraphic succession (Nisbet et al. 1977; Chapter 6), and it has proven possible to map stratigraphy everywhere that exposure is sufficient to determine rock types. In several localities such as along the Rupemba range, shears displace the stratigraphy in the Manjeri Formation but such displacements are limited to a few hundred metres and it is possible to match the displaced stratigraphic sequences across shear zones (within the limits of exposure). Further, Nd and especially Pb, isotopic compositions of some (but not necessarily all) Reliance Formation lavas are thought to reflect minor contamination by passage through much older continental crust (Chapter 7). It has not proved possible to map stratigraphic units within the rather monotonous lavas of the Zeederbergs Formation.

If it is supposed that the volcanic rocks of the Ngezi Group in Belingwe are indeed allochthonous, they must have been emplaced in a sheet 5-7 km thick and at least 30 km in width along a shear zone as thin as a few cm or less, and the shear zone must have been restricted throughout to exactly the same stratigraphic horizon above the Manjeri Formation. If such a Glarus-style tectonic emplacement took place it is surprising that subsidiary shears were not developed in previously serpentinised ultramafic rocks of the Reliance Formation. Feeder dykes to the volcanics have not been found cutting the Manjeri Formation, but dykes are seen only very rarely within the volcanic pile itself. Given the low viscosity of these lavas and their probable great areal spread from erupting rifts, this is not surprising. We take this as evidence that the volcanic feeders to the Ngezi Group are not exposed at the present erosion level.

We consider that the undeformed state of most of the rocks and the consistent facing directions and coherent stratigraphy mapped around most of the Belingwe Greenstone Belt can only be explained if the Ngezi Group is a conformable sequence, the basal unit of which is the Manjeri Formation. There is irrefutable evidence that this basal formation of the Ngezi Group greenstones was deposited unconformably on a granite-gneiss crust. If the Reliance Formation rests conformably on the Manjeri Formation, then by inference, the whole Ngezi Group was deposited above an older continental crust.

3.8 POST-GREENSTONE INTRUSIONS

3.8.1 *Ultramafic complexes*

Both the Chingezi and Shabani Gneisses as well as the greenstones are intruded by a number of layered ultramafic complexes. These are thought to belong to one suite (Mashaba Ultramafic Suite) related to the Mashaba Igneous Complex (Wilson et al. 1978). None of these intrusions has been dated but they post-date the ca. 2900 Ma Chingezi Gneisses and pre-date the ca. 2600 Ma Chilimanzi Suite adamellites. The Shabani Ultramafic Complex dips moderately towards the greenstone belt (Laubscher 1963) and this is interpreted to relate to D_g2 folding.

The Vukwe Ultramafic body is situated between the Brooklands Formation and the Shabani Gneiss. Most of this body comprises serpentinite with no relict texture. Layers of foliated amphibolite, similar to that in the Ngezi Schist inclusion, have been mapped particularly in the southern area. Specimens of dunite with relict fresh olivine, and pyroxenite with some relict clinopyroxene have been collected from the waste dumps of Kloof Mine. It is probable that the Vukwe Complex represents a differentiated ultramafic complex intruded into rocks of the Ngezi Schist inclusion. The age of the intrusion is unknown.

3.8.2 *Mashaba-Chibi Dykes*

These comprise two sets, an earlier set which trends east-southeast and a younger set which trends east-northeast. The dykes cut the Shabani Gneiss and the Shabani Ultramafic Complex east of the Belingwe Greenstone Belt and are in turn cut by the Shabani Younger Granite. To the east, in the area of the 3500 Ma Tokwe Gneiss, Wilson (1968) has shown that the trends of similar dykes (termed Mashaba-Chibi Dykes) form a radial pattern about the Mashaba Ultramafic Complex. Rare dolerite dykes possibly related to this set cut the Brooklands Formation and Martin (1983) describes a dyke of this suite cutting the Manjeri Formation.

Wilson (1968) records similar dykes south of Mashaba cut by the Chibi Batholith. The mineralogy of the dykes range from amphibolites (hornblende-plagioclase) to partly altered dolerites with relict hypersthene, augite and plagioclase. Some of the feldspar-poor dykes may have a komatiitic basalt composition.

3.8.3 *The Chilimanzi Suite: Chibi Batholith and related intrusions*

The youngest granitic intrusive rocks in the Belingwe area are the Chilimanzi Suite adamellites (Wilson et al. 1978). These include the Chibi Batholith which extends for 140 km with an approximate width of 20 km parallel to the margin of the Limpopo Belt south of the Belingwe Greenstone Belt, extensive areas of adamellite intruding the northern part of the greenstone belt and a number of smaller stocks.

The Chilimanzi Suite includes a range of compositions from granodiorite to true granite with a majority of adamellites. Textures range from fine grained to coarse and porphyritic and from massive to foliated. The rocks contain quartz-plagioclase-microcline with muscovite-biotite and more rarely hornblende in various proportions. Plagioclase is frequently partly altered. The Chibi Batholith has been described by Robertson (1974).

The intrusion of the Chibi Batholith truncates the main D_g2 syncline in the Ngezi Group greenstones and thus postdates D_g2. The deformation state of the Chilimanzi Suite intrusives varies. Undeformed massive textures are most common. The margins of the larger bodies and the smaller stocks are foliated. South of Brooklands the marginal phase of the Chibi Batholith comprises a finer grained rock which contains coarse strained and sutured quartz, partly replaced by a finer polygonal quartz, albite filled with inclusions of muscovite, biotite and epidote, unstrained myrmekite, strained potassium feldspar phenocrysts and unstrained microcline. Biotite appears to be recrystallised mimetically on a pre-existing fabric, with a few relict bent flakes. The rocks thus appear to have partly recrystallised at greenschist facies conditions (albite-microcline-biotite-muscovite-epidote). The foliation in the outcrops south of Brooklands strikes south-west similar to the D_g3 cleavage orientation in the adjacent greenstones (Fig. 3.17). South-west of Bend, Chilimanzi Suite plutons which intrude Chingezi Gneiss contain mafic xenoliths and these are variably flattened in an east-northeast striking plane. Foliation orientations do not appear to relate to position within small stocks and this deformation is correlated with D_g3 in the greenstone sequences. The mineral fabrics in the more strained plutons are interpreted as having recrystallised after D_g3 and this is consistent with fabric relations to pseudomorphs in the thermal aureole in the south of Brooklands described above.

Hickman (1978) has dated three plutons of the Chilimanzi Suite in the Fort Victoria (now Masvingo) area, which give a Rb-Sr whole-rock age of 2570 ± 25 Ma and an initial $^{87}Sr/^{86}Sr$ ratio of 0.704 ± 1. Two Rb-Sr whole-rock ages from the Chibi Batholith are consistent with this age but isotopic inhomogeneity between the suites implies that the Chibi Batholith contains components from more than one source region (Hawkesworth et al. 1979), or that Rb-Sr isotopes have been reset.

3.8.4 *The Great Dyke and East Dyke*

The Great Dyke (2460 ± 16 Ma, Hamilton 1977) and its parallel satellite dyke, the East Dyke (Worst 1956), both intrude the Belingwe Greenstone Belt. The Great Dyke cuts the Chingezi Gneiss and Tonalite and the north-west part of the greenstone belt. The East Dyke, trending north-northeast parallel to the Great Dyke, cuts the Chibi Batholith, the Brooklands Formation and the Shabani Gneiss Complex. It is displaced about 1 km by the Mtshingwe Fault and to the north of this can be traced as a series of isolated pods east-northeast along the southeast flanks of Rupemba Range, before it resumes its north-northeast strike just south of the Bannockburn-Rutenga Railway Line. It is probable that this change in orientation is related to the north-west striking D_g4 deformation as discussed in Section 3.6.3.

3.9 STRUCTURAL SUMMARY

The Shabani Gneiss Complex east of the Belingwe Greenstone Belt comprises migmatitic gneisses formed, complexly deformed and then intruded by more homogeneous granodioritic bodies in a relatively short period at about 3500 Ma ago. About 600 Ma later, at about 2900 Ma, to the west of the Belingwe Greenstone Belt, the Chingezi Gneiss Complex was likewise formed, complexly deformed and then intruded by a number of homogeneous plutons in a short period. At the same time the Shabani Gneiss Complex was intruded by the Mashaba Tonalite, which may, on the east of the belt, be the equivalent to the Chingezi Tonalite. Except for some shearing of the Mashaba Tonalite, the intense gneiss-forming deformation events identified in the Chingezi Gneisses are not recognized in the Shabani Gneiss Complex. The structural relations of the two areas of gneiss are not known; nor is the structural significance of the siting of the Belingwe Greenstone Belt on the boundary between these two gneiss provinces.

The Belingwe Greenstone Belt contains two sequences of greenstone, separated by an unconformity. The older Mtshingwe Group was possibly laid down on a basement of Chingezi Gneiss and infolded amphibolite in the west and on Shabani Gneiss in the east. Correlation of deformation events, clasts of possible basement tonalite in agglomerates and the contrast in metamorphic facies are all evidence of such basement to the

Mtshingwe Group in the west. In the east, correlation of deformation events, contrast of metamorphic facies and clasts of banded gneiss in breccias also point to deposition of the Brooklands Formation of the Mtshingwe Group on the adjacent Shabani Gneiss Complex.

The Mtshingwe Group was folded about east-northeast trending axes (D_g1) prior to the deposition of the younger Ngezi Group greenstone sequence. Basal unconformities between the overlying Ngezi Group and the underlying Shabani Gneiss Complex or the Mtshingwe Group are well exposed at four localities. Elsewhere this unconformity is confirmed by detailed mapping. Both greenstone sequences were then folded about a north-northwest axis (D_g2) which has preserved the Ngezi Group in a major syncline. Volcanism in the Ngezi Group is dated at 2692 ± 9 Ma (Chapter 7). A suite of east-west dykes (the Mashaba-Chibi suite) intrudes the Shabani Gneiss Complex, the Chingezi Gneiss Complex, the Hokonui Formation, the Shabani Ultramafic Complex and possibly the Brooklands Formation but has not been confirmed in the Bend Formation, nor above the Manjeri Formation in the Ngezi Group. Several mafic-ultramafic complexes were formed at about this time.

Intrusion of Chilimanzi Suite adamellites at 2570 Ma post-dates the D_g2 folding. D_g2 was followed by a further folding phase about north-east striking axial planes in the southern part of the greenstone belt and possibly north-west striking axial planes in the north (D_g3). Recrystallisation of the greenstone belt rocks to greenschist facies mineral assemblages, and partial retrogression of amphibolite facies assemblages to greenschist facies assemblages in the Shabani Gneisses, Chingezi Gneisses and the Chilimanzi Suite adamellites took place after the D_g3 deformation. Rb-Sr biotite-whole rock ages between 2500 Ma and 2460 Ma over much of the southern Zimbabwe Archaean Craton probably record the time of cooling or erosion after this event. Mineral ages in the most southerly part of this terrain of about 2000 Ma must reflect a Limpopo Belt event although no fabric evidence was recognised to substantiate this.

Intrusion of the Great Dyke and its satellite dyke, the East Dyke, at 2460 ± 16 Ma (Hamilton 1977) was followed by dextral displacement on the older Jenya Fault, north-south shearing on the Sabi Shear zone and dextral displacement on the Mtshingwe Fault, and some folding to the greenstone belt (D_g4).

3.10 INFERENCES

In general, our results confirm and amplify the conclusions of Coward et al. (1976), who studied the Northern Margin of the Limpopo Belt. From as long ago as 3500 Ma a varied granite-gneiss terrain has existed in the area. There is no direct evidence as yet available to show how this terrain was created – all that is known is that the very oldest gneisses have schist inclusions and even at 3500 Ma, a bimodal granite-greenstone terrain existed.

Deformation within the gneisses must have been intense. The (very poor) evidence available suggests that periods of granite intrusion were also periods of greenstone belt formation, within the 100 Ma resolution of available dating techniques. The first deformations of the older gneisses are not recorded, except as foliations prior to D_s1. Similarly, the oldest schist inclusions are probably only preserved as foliated amphibolite facies bands in gneiss. The Ngezi Schist inclusion, the oldest greenstone material to be preserved as an identifiable unit, is of poorly constrained age, probably ca. 3.5 Ga. Many other schist inclusions in the gneisses are assumed to be of similar age, and they may, in part at least, correlate with the Sebakwian rocks of this age preserved elsewhere on the craton (Wilson 1979).

Deformation and intrusion into the gneisses during and after the D_s1 and D_s2 events produced a basement to the craton. Later events such as the D_g1 event in the greenstone belt appear to have left little or no penetrative structural imprint on the gneisses but they warp the foliation trends on a kilometre scale at the present depth of erosion. The mafic dykes in the gneisses provide abundant evidence for regional tension at several periods, but it is difficult to correlate this with depositional events in the overlying crust. Possibly the mafic dyke intrusion between D_s1 and D_s2 was related to 3500 Ma Sebakwian volcanism; later the amphibolite facies metamorphism post-D_s2 may have been prior to uplift and erosion, cooling and then deposition of the Mtshingwe Group.

The stratigraphic, sedimentological and structural evidence implies that the Mtshingwe Group was laid down on top of this basement. In the east the Mtshingwe event has left little evidence in the gneisses, yet in the west of the belt a major event of roughly this age produced the Chingezi Gneiss and Tonalite. It is possible that the Mtshingwe event was a regional stretching of the basement, with marginal uplift (see Chapter 5). In the east the rapid facies changes in the Brooklands Formation imply that the margin of the depository was nearby, and that granite-gneiss terrain (with schist inclusions, as deformed mafic debris was also eroded) was being uplifted as the crust sank under what is now the greenstone belt. As subsidence progressed, volcanism began in the basin, and the Brooklands and Bend Formations were deposited on a continually subsiding basement. It is possible that D_g1 may reflect this. The deformation is recognized simply by the attitude of bedding – it has no associated cleavage. However D_g1 folded the foliation in the underlying gneisses on a large scale.

The first folding of the Ngezi Group greenstones (D_g2) does not appear to have been only a passive downwarping. Early shortening is recorded by a well developed cleavage in higher grade, deeper level rocks in the northern part of the belt. The geometry of the belt is that of a 'pinched' syncline. A hypothetical section is illustrated in Figure 3.18. The D_g2 deformation refolded the D_g1 syncline. It is not clear whether or not the Ngezi Group was deposited in a basin localised on the present outcrop of the greenstone belt (see Chapter 6); rather, the structural imprint of greenstone belt formation on the gneissic basement is uncertain. There is some evidence for regional tension in the gneisses during the formation of the volcanic suite of the Ngezi Group, since the east-west Mashaba-Chibi dyke suite cross-cuts all major units considered to be basement to the Ngezi Group greenstones, and also the contact of the Shabani Ultramafic Complex (Laubscher 1963) and perhaps one locality of the Manjeri Formation (Martin 1983). There is no direct evidence that this dyke suite intrudes higher Ngezi Group greenstones where dykes are extremely rare and of uncertain age.

The final major events, intrusion of the greenstones and basement by Chilimanzi Suite adamellite, the subsequent D_g3 folding and cleavage formation and the associated greenschist facies metamorphism are regional events.

CHAPTER 4

The Mtshingwe Group in the west: Andesites, basalts, komatiites and sediments of the Hokonui, Bend and Koodoovale Formations

J.L.ORPEN, A. MARTIN, M.J. BICKLE & E.G. NISBET

ABSTRACT

The Mtshingwe Group constitutes the lower greenstones of the Belingwe Belt. It includes the Hokonui, Bend and Koodoovale Formations in the west as well as the Brooklands Formation in the east of the belt.

Most of the exposure of the Hokonui Formation is in the region to the north and south of the town of Mberengwa. It comprises a suite of mafic, intermediate and felsic volcanic rocks (dominantly the latter two categories), together with minor sedimentary rocks.

The Bend Formation is a thick suite of mafic, ultramafic and tholeiitic volcanic rocks, most of which are clearly extrusive, and which appears to lie unconformably upon the southern flank of the Hokonui Formation. It also includes minor but prominent banded ironstones and minor felsic volcanic rocks. Many of the komatiites display superb crystallisation textures.

The Koodoovale Formation lies unconformably but concordantly upon the Bend Formation, and is itself overlain with major unconformity by the Ngezi Group. It includes a conglomerate and a felsic agglomerate member.

4.1 INTRODUCTION

The Mtshingwe Group includes in the east the Brooklands Formation, and in the west, the Hokonui, Bend and Koodoovale Formations. In this chapter the latter three are described, in the next the Brooklands Formation is discussed, since it is not anywhere in contact with the other three formations and is classified with them only on structural style and position.

The three western formations of the Mtshingwe Group form an impressive suite of Archaean strata (Fig. 4.1): The Hokonui Formation is essentially a complex andesitic volcano; the Bend Formation may be the product of extensional mafic to ultramafic volcanism; and the conglomerates and felsic agglomerates of the Koodoovale Formation postdate uplift and some erosion but not folding of the Bend Formation. Contacts between the formations all appear to be unconformable, but there is no strong evidence for major folding between the deposition of the upper two formations, and it is possible that the Hokonui pile was also essentially undeformed when subsidence occurred on its flanks to form a base for the Bend volcanism.

69

70

Figure 4.1. a) Slopes of Belingwe Peak from the south; mainly Bend Formation. b) Hand specimen of Clifton Tuff GR 061302.

4.2 HOKONUI FORMATION

4.2.1 *Introduction and structural setting*

The Hokonui Formation comprises a suite of mafic, intermediate and felsic volcanic rocks and minor sedimentary rocks. It takes its name from the Hokonui hills of South Island, New Zealand, via the ranch of Senator Garfield Todd. It is dominantly extrusive with a few mafic/ultramafic intrusions. It outcrops between the Chingezi Gneiss terrain to the west and the Bend Formation and the overlapping Manjeri Formation to the east. Exposure over much of this area is poor, good exposures being restricted mainly to the Mtshingwe and Dohwe rivers and nearby areas. Over much of the area, particularly in the north and west, mapping of mafic and felsic units is based only on soil colour. Cleavage is variably developed (e.g. Fig. 3.16) and although many outcrops are well cleaved, some less deformed outcrops show well preserved examples of all the main rock types.

At its base the Hokonui Formation is thought to be intruded by the Chingezi Tonalite and either in unconformable or gradational contact with the foliated and isoclinally folded amphibolites of the Bvute Formation. Possible interpretations are discussed in Chapter 3. The Hokonui Formation is overlain with apparent conformity by the Bend Formation although poor exposure on this contact makes a definite interpretation unsafe and the contact is probably faulted. Koodoovale conglomerates (see below), which disconformably overlie the Bend Formation, contain clasts of Hokonui-like felsic volcanic detritus. The Manjeri Formation has a well exposed low-angle unconformable contact with Hokonui Formation felsic volcanic rocks in the Mtshingwe River and lies with an exposed and unambiguous unconformity on Bend Formation rocks (Chapter 6). Substantial erosion of Bend Formation and the upper part of the northern outcrop of the Hokonui Formation prior to deposition of the Manjeri Formation is inferred. The Hokonui Formation pinches out to the south between the Bend Formation and the Bvute Formation but whether this is a primary volcanogenic feature, an erosive or a tectonic feature is indeterminable. In the north, the Hokonui Formation is cut out by granite of the Chilimanzi Suite. The internal structure of the Hokonui Formation is also unresolved. In the better exposed sections it comprises alternating mafic and felsic rocks. In many more poorly exposed areas even strike is not well defined and parts may be repeated by folding. The apparent thickness in outcrop of the formation is up to 8 km but its true stratigraphic thickness may well be less than this.

4.2.2 *Rock types – Mafic and felsic rocks*

Mafic rocks form a significant part of the Hokonui succession particularly in the western lower parts of the succession. There is considerable uncertainty as to whether higher grade amphibolites near the base of the formation are transitional to, or part of, the Bvute Formation. However, higher in the succession both fine-grained and coarser-grained ophitic-textured low grade mafic units are interbedded in the felsic rocks. The fine-grained mafic rocks are recrystallised to greenschist assemblages of tremolite/actinolite, inclusion-rich plagioclase, clinozoisite, chlorite and opaques, with variable degrees of carbonation. No volcanic structures are preserved. The coarser grained mafic rocks display ophitic textures with augite partly replaced by tremolite enclosing cloudy plagioclase with bastite pseudomorphs (after orthopyroxene?), chlorite and clinozoisite/epidote. The mafic rocks are interpreted as lavas or sills.

The light-coloured felsic volcanic rocks are distinctly plagioclase-phenocryst rich and were deposited as a range of fragmental units. Fine-grained graded and ripple-bedded

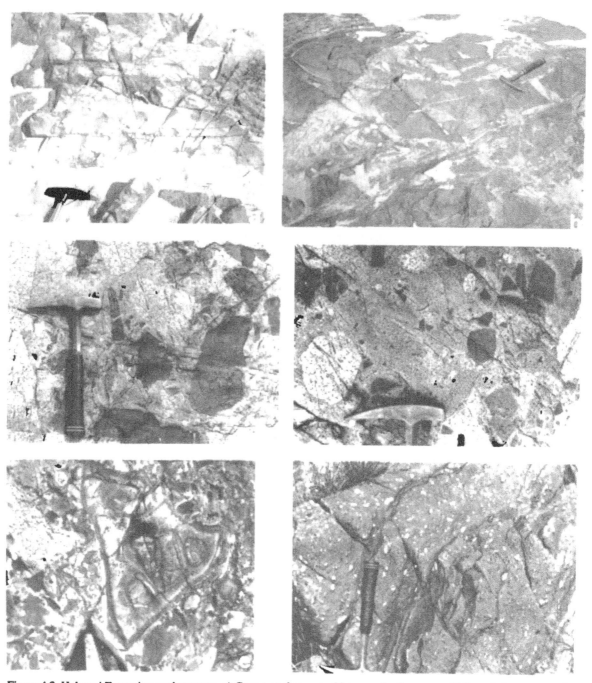

Figure 4.2. Hokonui Formation agglomerates. a) Coarse agglomerate: Note angular fragments. b) Breccia in coarse agglomerate. c) Detail of coarse agglomerate, showing tuff clasts (dark) and tonalite clasts (light). d) Detail: Note rounding of some clasts. e) Alteration halo around pyroclast. f) Felsic clast agglomerate.

tuffs, some with flame structures, lapilli tuffs and coarser volcanoclastic rocks with up to 20 cm angular or sub-angular clasts are found (Figs 4.1b, 4.2). Clasts contain abundant inclusion-rich plagioclase phenocrysts, now albite, and rare clinopyroxene phenocrysts, set in a fine grained groundmass of feldspar microlites, quartz, chlorite and clinozoisite. The matrix to the clasts has a similar mineralogy. Secondary carbonate minerals are frequently abundant. Crystal tuffs with abundant 2 mm plagioclase megacrysts occur near the top of the formation. Rare exotic clasts including chert, grey porphyritic, granitic and mafic fragments are found in the fragmental felsic rocks. More massive phenocryst-rich rocks are exposed in the Dohwe river 2.2 km southwest of the Mtshingwe River confluence. These may represent lavas. In composition the felsic rocks range from andesite to dacite (Chapter 8). No mafic rocks were analysed.

A distinctive volcanic breccia is well exposed in the Mtshingwe River 4.5 km up-stream of the Bulawayo-Shabani road bridge. This contains blocks of a tonalite up to 10 m in diameter set in a matrix of smaller felsic volcanic clasts and fine grained felsic matrix (Fig. 4.2). The tonalite clasts are identical to the Chingezi Tonalite and this outcrop is interpreted as a volcanic vent although poor exposure in the surrounding area precludes a proper study of its field relations.

Felsic, probably mainly intrusive rocks, are well developed, if poorly exposed around the Belvedere Mine. These contain quartz phenocrysts, and some plagioclase pheno-crysts in a fine grained quartz-feldspar and minor chlorite, epidote and muscovite. Some of these rocks are relatively coarse grained leuco-tonalite. The felsic dykes cutting the Bvute Formation are tentatively correlated with the Hokonui Formation. They contain plagioclase, corroded quartz and pseudomorphs filled with muscovite and haematite, and K-feldspar as phenocrysts in a fine grained groundmass of quartz and plagioclase. One outcrop of an identical rock type was observed within the Hokonui Formation (Orpen 1978).

4.2.3 *Sedimentary rocks*

Sedimentary rocks form a very small proportion of the Hokonui succession. A few chert-ironstone horizons are exposed near the top of the succession. At the Belvedere gold mine thin horizons of chert and sulphide facies ironstone (pyrite, pyrrhotite and ar-senopyrite) are associated with graphitic argillite. Elsewhere a quartz-sericite-fuchsite schist, an arkose and quartzite have been noted.

4.2.4 *Mafic-ultramafic intrusions*

The Vanguard ultramafic sill, the Knott pyroxenite, the Gurumba Tumba ultramafic sill and dykes of the Mazvihwa suite all cut the Hokonui Formation. The relatively well preserved Vanguard sill and Gurumba Tumba ultramafic are included in the Mashaba ultramafic suite on the relatively slender evidence of their similarity in deformational state.

4.2.5 *Interpretation*

The Hokonui Formation comprises a suite of felsic, intermediate and subordinate mafic volcanic rocks occupying an outcrop area up to 8 km across strike and 35 km in the strike length. The Formation pinches out in the south but is cut out just to the north of its widest outcrop by younger granite. These outcrop dimensions are similar to modern andesitic volcanoes and to felsic volcanic 'centres' identified in the 3500 Ma Pilbara greenstones

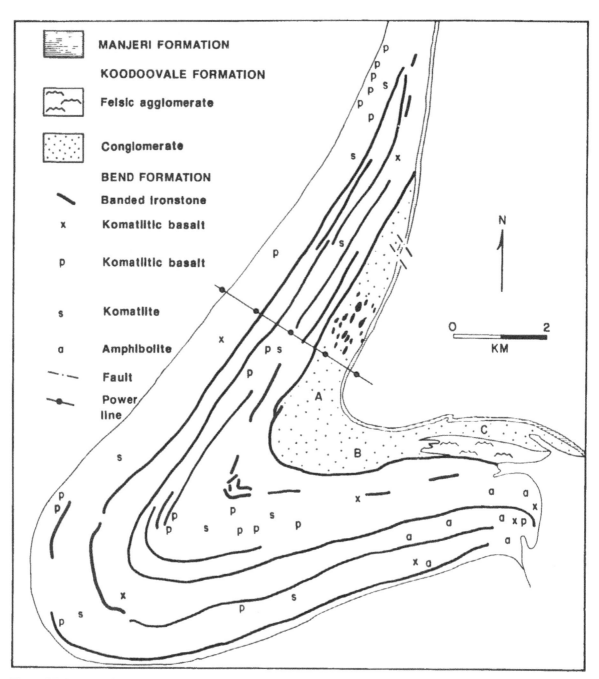

Figure 4.3. Bend and Koodoovale Formations: Outcrop distribution of conglomerates facies (solid) within phyllite illustrated only north of power line.

74

(Barley et al. 1979) and in the Canadian Archaean (Dimroth et al. 1981). Poor exposure precludes a proper interpretation of the setting of this unit but the presence of a coarse (vent?) breccia in the thicker part of the outcrop is consistent with this being a predominantly andesitic volcanic pile centred in the region of the Mtshingwe River with the pinch-out below the Bend Formation to the south being a primary volcanogenic feature. Part of the succession may have been laid down subaerially.

4.3 THE BEND FORMATION

The Bend Formation (Figs 4.1, 4.3) includes a thick suite of mafic, ultramafic and tholeiitic volcanic rocks, most of which are extrusive, with intercalated banded ironstones and minor felsic volcanic rocks. It lies with apparent unconformity on the southern flanks of the Hokonui Formation and is itself overlain conformably by the Manjeri Formation of the Ngezi Group, or elsewhere by the Koodoovale Formation. The formation is capped by a layer of banded ironstone at the top of the suite of high magnesium lavas. The ironstone is remarkably thick, up to 100 m. It is separated by a disconformity from the overlying Koodoovale Formation. The Bend Formation has been folded around the Peak syncline and Dube anticline, which has tectonically thickened the strata to six kilometres along the synclinal axis. Throughout the Formation younging directions are plentiful; all are towards the centre of the syncline indicating a continuous succession of lavas without apparent tectonic repetition (Fig. 4.4).

The following description outlines the texture and petrography of the volcanic rocks in the formation. Detailed geochemical discussion is left to Chapter 8.

4.3.1 *Field relationships*

The relationship between the Bend and Hokonui Formations is not clearly expressed and the contact is poorly exposed. Orpen (1978) has reported evidence for deformation in the Hokonui Formation prior to deposition, and for westward-younging of the Hokonui Formation below the Bend Formation. Martin (1983) found no similar evidence to the north, however. The wedging out of the Hokonui southwards and the apparent overstepping of the Bend onto the amphibolites of the Bvute Formation in the keel and southern limb of the Peak Syncline might be explained by an unconformity: The contact could also be a conformable contact over an aerially limited calc-alkaline volcanic centre or a tectonic contact.

A period of erosion followed the deposition of the Bend Formation prior to the accumulation of the thick sedimentary and felsic volcanic pile of the Koodoovale Formation. Three wash-out channels found in the banded ironstone of the Belingwe Peak Mountain at the base of the Koodoovale Conglomerate provide evidence of this erosional break, but as exposed it is not possible to demonstrate a regional unconformity between the two formations.

Between the Bend Formation and the Ngezi Group the contact exposed to the north of the Koodoovale Formation clearly displays a truncation of the older banded ironstone.

4.3.2 *Petrography*

The stratigraphic succession in the formation is detailed schematically in Figure 4.4. A set of discrete volcanic events can be identified, each of which includes an assortment of

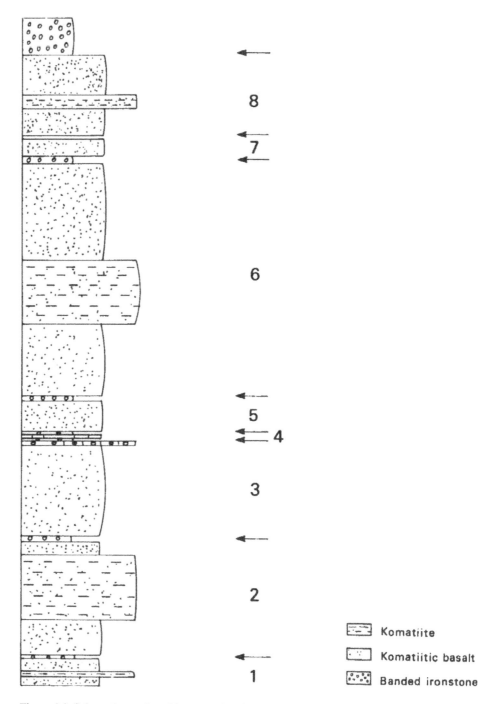

Figure 4.4. Schematic stratigraphic succession in the Bend Formation. Numbers denote volcanic sequences separated by banded ironstone horizons.

Legend:

- Komatiite
- Komatiitic basalt
- Banded ironstone

mafic to komatiitic lavas, and each event being bounded by a period of volcanic quiescence marked by banded ironstone deposition.

The formation consists for the most part of massive and pillowed basalts and komatiitic basalts, clinopyroxene spinifex-textured komatiitic basalts and olivine spinifex-textured komatiites associated with thick (komatiitic) peridotite units (Fig. 4.4). Thin 2 to 30 m oxide-facies banded ironstone and chert horizons are interbedded with the volcanic rocks. Although the Bend Formation underlies some of the most rugged terrain in the Belingwe area only the chert and ironstone horizons are well exposed. Ten horizons are mapped in the core of the syncline and some of these can be followed along the entire 20 km strike length of the formation (Figs 4.3 , 4.4).

Outcrops of the volcanic rocks are patchy and continuous sections across individual units are rarely exposed. The volcanic structures and igneous textures are generally well preserved although pillows are often somewhat flattened. Banded ironstone and chert horizons are exposed in contact with underlying massive or pillowed mafic rocks and more rarely in contact with olivine spinifex-textured komatiites. The laterally continuous banded ironstone horizons would appear to have been deposited during intervals between eruption of the volcanic rocks. Within some better exposed volcanic units, basalts and komatiitic basalts predominate towards the base and komatiites towards the top (Fig. 4.4). The continuity of individual ironstone-chert horizons suggests that the Bend Formation has not undergone major internal disruption although strain during folding must have been taken up on shear zones. Horizons of highly sheared mafic rocks occur sporadically throughout the sequence, one such horizon being particularly well developed below the second banded ironstone horizon above the base of the Bend Formation.

4.3.3 Metamorphism and alteration

All the volcanic rocks within the Bend Formation show some degree of recrystallisation to greenschist facies mineral assemblages. Unaltered clinopyroxene is commonly preserved in both the komatiitic basalts and the komatiites. The least-altered mafic volcanics contain fresh clinopyroxene in a fine grained groundmass after glass. The more altered basalts have recrystallised to tremolite-epidote-chlorite-albite-sphene-quartz assemblages although igneous textures (mainly pseudomorphs of clinopyroxene or plagioclase) are commonly preserved. The komatiites and peridotites exhibit a similar range of preservation states with well preserved pseudomorphs after olivine filled with serpentine, magnetite and more rarely chlorite or tremolite. Quench clinopyroxene and skeletal chromites are preserved in a groundmass reminiscent of glass. In two of the more altered samples the groundmass is recrystallised to an aggregate of fine grained serpentine, tremolite and chlorite.

Ironstone horizons are fine grained (ca. 0.01 mm) and contain assemblage ranging in grade to the maximum phase assemblage quartz-magnetite-haematite-stilpnomelane. Grunerite has been found in one horizon in the underlying Hokonui Formation.

4.3.4 Komatiites

The komatiites occur as units between a few and fifty metres in thickness. These units appear to constitute parts of typical spinifex and cumulate textured layered flows (Fig. 4.5) with thick lower cumulate zones enriched in phenocryst olivine and a zoned upper spinifex unit. However poor exposure makes it difficult in places to relate isolated outcrops of spinifex-textured komatiite to their basal cumulate zones. The best exposed cross-section of such a flow is illustrated in Figure 4.5 and this typifies the variation in

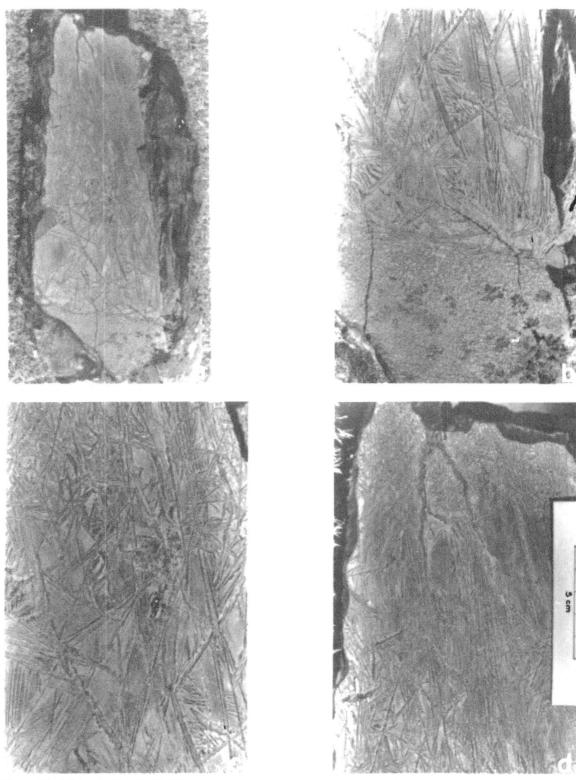

Figure 4.5. a) Cross-section of a spinifex-textured unit as a komatiite flow (for scale see Fig.4.5d). b) Detail of Figure 4.5a. Base of unit, with cumulate base, foliated spinifex zone and randomly orientated pseudomorphs after olivine spinifex. c) Detail of Figure 4.5a. Middle of unit, showing randomly orientated pseudomorphs after olivine defining domains in which books of smaller plates grew.d) Detail of Figure 4.5a. Top of unit, with sheaves of orientated pseudomorphs after olivine plates growing down from the flow top.

type of spinifex texture. Most of the spinifex-textured samples contain parallel sets of 0.01 to 1.0 mm thick plates of olivine now pseudomorphed by serpentine and magnetite. The plates range up to tens of centimetres in length.

In thin section (Fig. 4.6) the flows characteristically have cumulate bases which contain equant, non-skeletal olivine pseudomorphs 0.5-2 mm across. Most pseudomorphs are rounded and the packing of the crystals is usually close with an interstitial groundmass after glass containing clinopyroxene microlites with both skeletal and arborescent textures. Skeletal chromite in fine snowflake-shaped grains occur. The proportion of glass increases rapidly towards the top of the flow. Typical crystal forms are shown in Figure 4.6. Figures 4.6a-c show cumulate textures, while Figures 4.6d-l show skeletal and spinifex textures.

Amygdales are rare in the Bend komatiites and none has been found over 1 mm in diameter (Fig. 4.6m). They occur in both cumulate and spinifex textured units. In the spinifex units amygdules appear to be associated with a peculiar vermicular olivine texture (Fig. 4.6n). Clinopyroxene crystallites in the groundmass are generally acicular and typically consist of thin-walled augite prisms enclosing a core of altered glass or serpentine. The textural variety in the clinopyroxene laths is considerable (Figs 4.6o-r). Finally, some spinifex units have skeletal magnetites (Fig. 4.6s).

4.3.5 *Komatiitic basalts and basalts*

The komatiitic basalts and basalts exhibit characteristic textures. Textural types, including macroscopic clinopyroxene spinifex textures, microscopic skeletal or hollow clinopyroxene often with very fine fan-like (arborescent) clinopyroxene in the groundmass and samples with equant clinopyroxene phenocrysts. Figure 4.7 illustrates some typical textures.

The clinopyroxene spinifex-textured rocks are similar to those described in other areas (e.g. Nisbet et al. 1977; Arndt et al. 1977) and include flows with clinopyroxene needles up to 40 cm in length. Exposure is insufficient to map complete sections through such flows. The majority of the other samples from both the massive and pillowed units exhibit skeletal clinopyroxene or fine arrays of arborescent clinopyroxene in the groundmass. The skeletal clinopyroxene grains often have chlorite-filled cores. One group of five rocks collected from adjacent outcrops of a distinctly 'spherulitic' pillowed lava contains equant or more rarely hollow (chlorite-filled) clinopyroxene micro-phenocrysts in a matrix recrystallised to actinolite-chlorite-epidote-albite-quartz and rare calcite. This group of rocks plots in the tholeiite field on the basis of Al_2O_3 to $FeO/(FeO + MgO)$ ratios as discussed in Chapter 8, although rocks with similar textures plot in the komatiite field. 'Spherulitic' textures or ocelli are common in both basalts and komatiitic basalts. Light coloured spherical or irregular and partly coalesced areas are set in a darker matrix.

4.3.6 *Banded ironstones*

The banded ironstone horizons are well-exposed and many may be traced continuously over long distances. The second horizon from the base is exposed for 20 km along strike; the 100 m thick horizon forming the mountain peak is 12 km long. Many other bands, from 2-30 m thick, can be traced for some distances with little thickness variation. The rock is typical of oxide-facies ironstone and no signs of sulphide or carbonate facies were seen. The fine bands are seldom more than a few mm thick and fine detail of apparently primary origin is preserved in some outcrops. The rocks are dominantly magnetite and chert with some haematite. Minor stilpnomelane occurs locally.

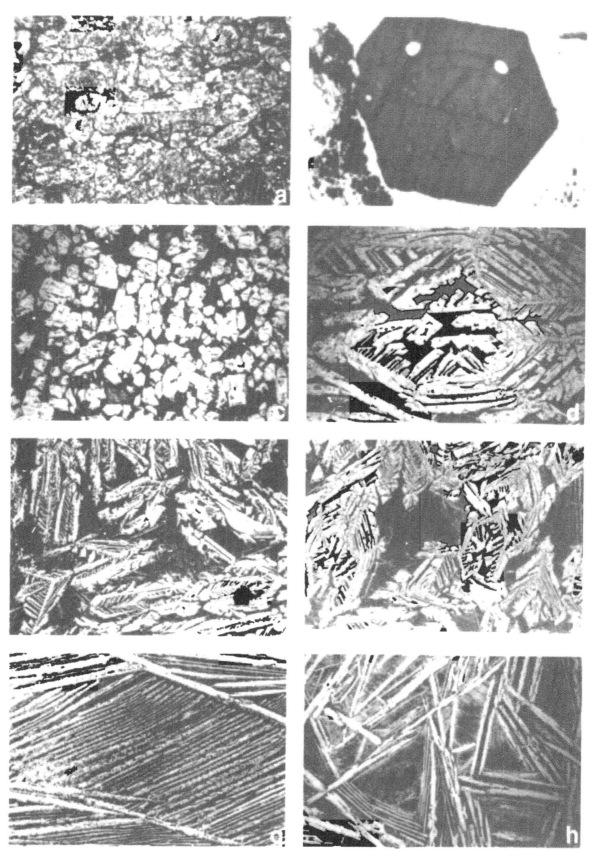

Figure 4.6. Typical crystal froms. See also next pages.

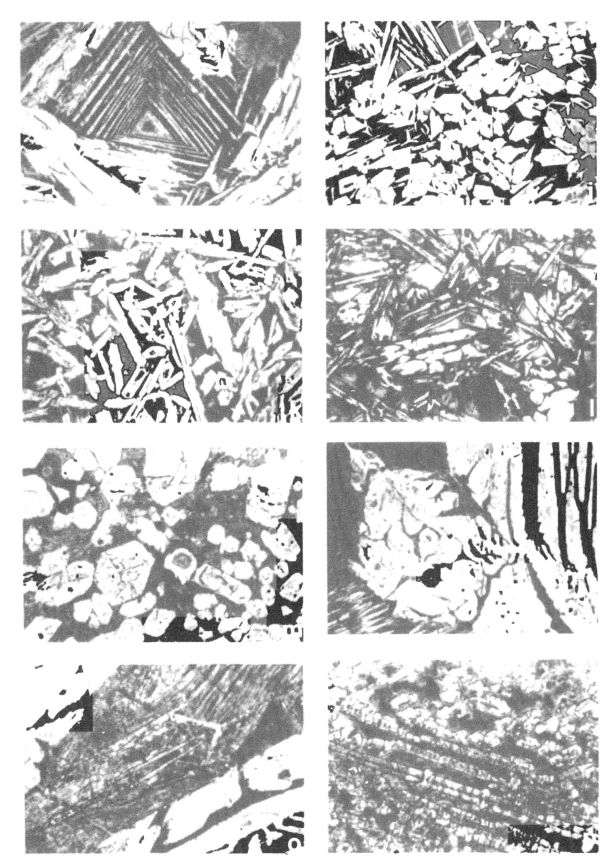

Figure 4.6. Typical crystal forms. See also previous and next page.

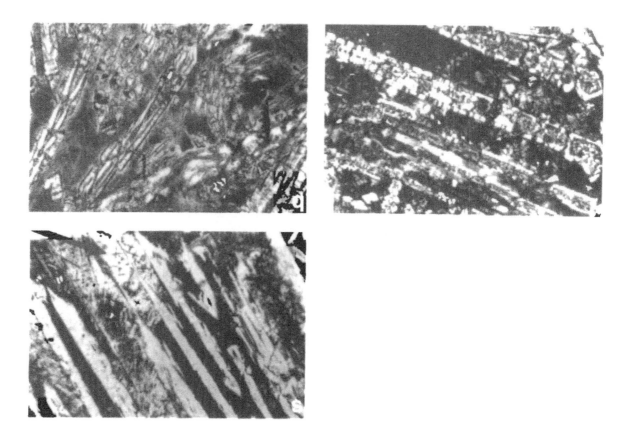

Figure 4.6. Typical crystal forms. See also previous pages. a) Cumulate base of komatiitic flow Bend Formation. Non-skeletal pseudomorphs after olivine crystals in a once-glassy matrix (specimen SL 511). Field of view 8 mm wide, crossed polars. b) Euhedral chromite in komatiite flow (specimen SL 835). Field of view 0.9 mm wide, open polars. c) Cumulate base of komatiite flow near spinifex unit. Note increase in proportion of once-glass and sharp faces of olivine pseudomorphs (specimen SL431). Field of view 8 mm wide, crossed polars. d) Highly dendritic olivine growth from a central septum (specimen SL722). Field of view 8 mm wide, open polars. e) Spinifex texture in olivine pseudomorphs (specimen SL724). Field of view 37 mm wide. Open polars. f) Olivine spinifex texture, now pseudomorphed (specimen SL722). Field of view 37 mm wide, open polars. g) Olivine spinifex texture, now pseudomorphed (specimen SL722). Field of view 37 mm wide, open polars. h) Olivine spinifex texture, now pseudomorphed (specimen SL722). Field of view 37 mm wide, open polars. i) Olivine spinifex texture, now pseudomorphed. 'Books' of skeletal olivine occur in the interstices of randomly orientated crystals; the large crystals define domains controlling the growth of smaller crystals (specimen SL446). Field of view 8 mm wide, open polars. j) Pseudomorphed olivine spinifex texture. Small skeletal crystals (specimen SL741A). Field of view 8 mm wide, open polars. k) Pseudomorphed olivine spinifex texture. Small dendrites (specimen SL445). Field of view 8 mm wide, open polars. l) Pseudomorphed olivine spinifex texture. Minute skeletal olivine pseudomorphs in altered glass (specimen SL329). Field of view 1.3 mm wide, open polars. m) Amygdule in komatiite (specimen SL569). Field of view 3.4 mm wide, open polars. n) Vermicular-granular olivine texture (specimen SL721A). Field of view 3.4 mm wide, open polars. o) Interstitial augite plating a pseudomorphed olivine dendrite in komatiite (specimen SL721B). Field of view 3.4 mm wide, open polars. p) Skeletal clinopyroxene in komatiite (specimen SL721B). Field of view 1.3 mm wide, open polars. q) Skeletal clinopyroxene in komatiite (specimen SL721B). Field of view 1.3 mm wide, open polars. r) Multi-lamellar twinning in skeletal clinopyroxene in komatiites (specimen SL722). Field of view 3.4 mm wide, open polars. s) Magnetite dendrites in komatiite. Field of view 3.4 mm wide, open polars.

4.3.7 *Felsic rocks*

A horizon of reworked felsic clasts occurs in the formation. The clasts may have been derived from an eroded terrain of the Hokonui Formation or of similar material. Clasts of spinifex-textured basalt also occur.

The massive felsic sills have a granophyric texture and contain small plagioclase phenocrysts.

4.4 THE KOODOOVALE FORMATION

The Koodoovale Formation contains the uppermost part of the lower greenstone Mtshingwe Group sequence. It lies unconformably on the Bend Formation which is locally eroded, and is itself unconformably overlain by the basal Manjeri Formation of the Ngezi Group. The Koodoovale Formation is approximately 1 km thick, most of which is fluviatile conglomerate. It is folded around the Peak Syncline, and at the eastern end of the southern limb of the syncline the conglomerate interdigitates with 500 m of felsic volcanic rocks. The formation is thus divided into a conglomerate and a felsic agglomerate member (Fig. 4.3).

4.4.1 *Conglomerate member*

The conglomerate (Fig. 4.8) is typically a polymict rock, with clasts of granite, porphyritic felsites, dark-green tremolite-chlorite schists, grey-green basalts (sometimes with spinifex-textured clasts) some dolerite and common banded ironstone fragments. Only locally do oligomict channel infills occur (Fig. 4.8b). In general, clasts are very well rounded and have a high degree of sphericity, especially in areas B and C on Figure 4.3 (Fig. 4.8). In area A clasts average about 15 cm long, with occasional cobbles to 20-25 cm in long axis. In area B clasts range up to 50 cm. The rock is massive and poorly sorted; bedding is difficult to identify, although sand lenses rarely occur in area B. The matrix is a coarse sand with a high chlorite content.

In area C (Fig. 4.3) the facies is somewhat different. Most clasts are porphyritic felsite, banded ironstone and the general lithological diversity of clasts is much less than in the other areas. Bedding is frequently expressed in alternating layers of coarse conglomerate and fine sand. Sand beds are from a few cm to 30 cm thick and sometimes show grading. Facing directions are always to the north, into the syncline. Clasts in the conglomerate beds range up to 20 cm across: The whole conglomerate beds may be several metres or more thick.

4.4.2 *Channelling in the conglomerate member*

The Koodoovale Conglomerates show several interesting outcrops displaying erosive channels. One of the best of these is on the top of the mountain ridge at GR 026187 where the cross-section of a large channel is exposed. Figure 4.9 shows a detailed outcrop map. The 450 m wide channel is asymmetric in shape, with a steep northern bank 60 m high, while the opposite side slopes gently down to the deepest point 70 m from the base of the cliff. The banded ironstone cap of the Bend Formation is 80 m thick at this point along the mountain and the channel almost eroded completely through it to the underlying basalt.

Outcrops of conglomerate in the deeper part of the channel near the cliff bank are noticeably coarser and less sorted than those further south. All the clasts are well-rounded

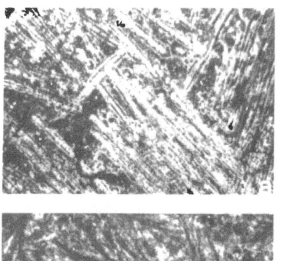

Figure 4.7. Textures in Bend Formation komatiitic basalt and basalt. a) Clinopyroxene spinifex texture in komatiitic basalt. Field of view 8 mm wide, crossed polars. b) Skeletal clinopyroxene prisms in komatiitic basalt. Field of view 8 mm wide, open polars. c) Plumose augite in tholeiite. Field of view 8 mm wide, open polars.

with mean diameters between 5-10 cm, but up to 30 cm across. The clasts are variable in composition, including felsites, basalts and granites, although locally banded ironstone slightly predominates. A thin four-metre thick banded ironstone horizon overlies part of the channel.

In general, the contrast between the non-clastic nature of the topmost Bend Formation ironstone and the thick coarse Koodoovale Formation is a striking record of a nearby major uplift. The generally polymict nature of the clasts, including granitic material, implies regional uplift. Some clastic material was probably locally derived from the Bend and probably the Brooklands Formation, but there must have been a large scale uplift of the regional granite-gneiss terrain, as well as of the Hokonui Formation terrain. The long strike continuity, textures and lack of lateral variation in the Bend Formation implies a flat-floored subaqueous setting with little topographic relief: In contrast the Koodoovale Formation shows erosional features and facies suggesting that it was probably deposited subaerially for the most part, as part of a large uplifted terrain. Relief must have been considerable to produce the 1 km thickness of the Koodoovale.

4.4.3 *Felsic agglomerate member*

The Felsic member is poorly exposed and no contacts are exposed between it and the conglomerate member. In one locality a thin tuff horizon appears to lie between the two: in another a banded ironstone occurs at this level. The clasts are for the most part rhyolitic to dacitic in composition and often heavily carbonated. They are composed of quartz phenocrysts, plagioclase, rare biotite and clinopyroxene and variable amount of calcite. The matrix of most clasts is felsic with a low chlorite content.

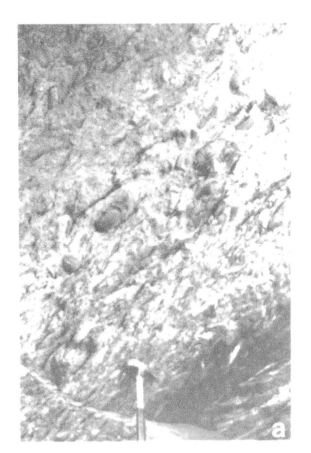

Figure 4.8. a) Koodoovale conglomerate, showing cleavage and flattening of clasts. b) Local oligomict banded-ironstone conglomerate from small channel (GR 051174) in area B. c) Rudaceous bed, area C, showing sphericity of clasts (GR 085180).

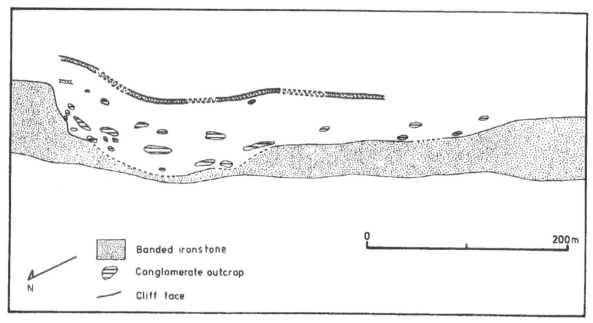

Figure 4.9. Channel at the base of the Koodoovale Formation cut into the banded ironstone capping the Bend Formation (GR 026187).

4.5 CONCLUSION

The Mtshingwe Group is thought to have been deposited on a slightly older gneissic crust (Chapter 3). It contains three formations deposited in markedly distinct sedimentary environments. A thick and localised andesitic volcanic pile, the Hokonui Formation, was succeeded by the stratigraphically continuous komatiite-ironstone succession of the Bend Formation. It is probable that these formations were formed in rather different tectonic environments. The overlying Koodoovale Formation characterized by polymict conglomerate is a 'molasse'-like deposit probably formed during the major folding and uplift of the Mtshingwe Group.

A possible modern analogy with the western Mtshingwe Group is the region to the south and west of Rotorua, New Zealand. Here, there are several andesitic to felsic centres, such as Mt. Tarawera, together with active hydrothermal systems and a major rift graben. It is possible that, in the Mberengwa area, the development of the Hokonui volcanic edifice was followed by rifting, subsidence, and the evolution of a major graben filled with komatiite and komatiitic basalt lava and ironstone. Later, uplift on the flanks of the graben provided detritus for the Koodoovale conglomerates, and tilting of the Bend Formation.

CHAPTER 5

Sedimentology of the Brooklands Formation, Zimbabwe: Development of an Archaean greenstone belt in a rifted graben

E.G. NISBET, M.J. BICKLE, A. MARTIN & J.L. ORPEN

ABSTRACT

The Brooklands Formation, part of the ca. 2800-2900 Ma Mtshingwe Group, contains a wide variety of sedimentary and volcanic rocks. These include coarse breccias, thought to have been laid down by debris flows or avalanches; conglomerates, siltstones and shales, and banded ironstones, as well as komatiitic basalt and komatiite flows. The sequence was probably laid down on 3500 Ma old gneissic crust, at one margin of a late Archaean graben.

5.1 INTRODUCTION

The Brooklands Formation, part of the Mtshingwe Group (Fig. 5.1) contains a wide variety of sedimentary and igneous rocks formed at the inception of a late Archaean (possibly 2900-2800 Ma old) greenstone sequence. The formation contains both igneous and sedimentary rocks, with very extensive lateral facies changes (Fig. 5.2). The sequence is thought to be conformable, and to have been laid down on older gneissic terrain. It has been metamorphosed to greenschist facies (Chapter 3).

Deformation and weathering have obscured many sedimentary structures, which tend only to be visible in the better preserved of the river sections. This considerably hampers the sedimentological study of regional facies changes: Even information about current directions is very difficult to obtain. Furthermore, all sedimentary structures which are preserved have been rotated by deformation. Objects such as boulders in conglomerates have been considerably strained by three deformation phases.

In the following description sedimentological terminology is used throughout, but it should not be forgotten that these are metamorphic rocks. However, the formation preserves sufficient sedimentological features that a detailed sedimentological analysis is justified.

5.2 STRATIGRAPHY

Four distinct members may be distinguished. The two mainly sedimentary members increase significantly in thickness across the Mtshingwe Fault (Figs 5.1, 5.2).

87

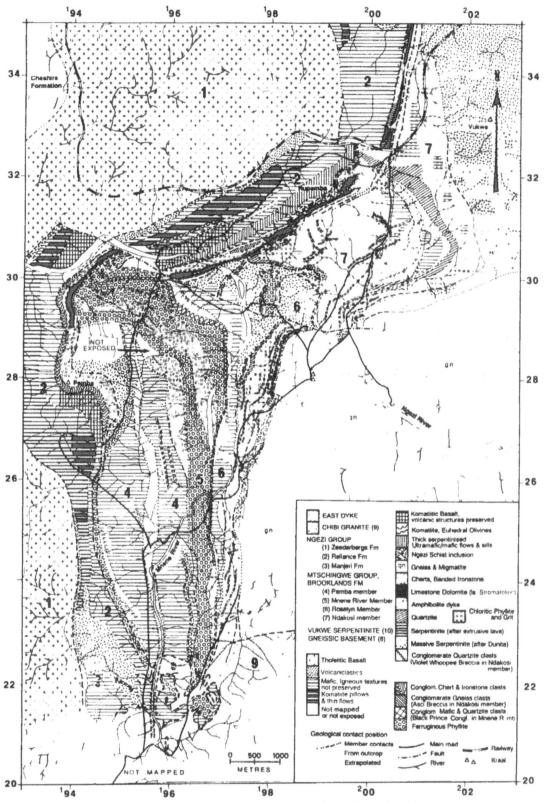

Figure 5.1. Map of the Brooklands Formation, Belingwe. Mapping by M.J.B. and E.G.N., 1974 and 1975. Grid from topographic 1:50 000 sheet, Mnene 2030 C1, edition 2, 1974.

5.2.1 Ndakosi Member

The dominant rock types in the Ndakosi Member are quartz grits, siltstones and pelites, which are interbedded with quartzites, conglomerates, breccias, cherts and banded ironstones. Associated mafic and ultramafic bodies may be intrusive. Predominant minerals are chlorite-muscovite-quartz. Igneous rocks show no obvious extrusive features and

Table 5.1. Ndakosi Member.

Height in section (m)	Type section of Ndakosi Member	Regional variation and comments
	Conglomerate breccias, clasts of banded gneiss (Asci beds) pelite, chert and banded iron formation	Not present in type section, enclosed in serpentinite probably derived from dunite intrusives
2495		
	Possible felsic volcanic rocks 1-3 cm rounded felsic clasts in felsic matrix	
1995		
	Pelites, chert bands, quartz pebble bands	Limited lateral extent, thickness 0-200 m (enhanced by D_g3 strain)
1595		
	Quartzite breccia (Violet Whoopee beds)	Limited lateral extent and rapid lateral changes in thickness and clast size (maximum quartzite clast 3.0×0.5 m)
1570		
	Pelites, chlorite rich siltstones and grits	
1330		
	Massive mafic, igneous structures not preserved	
1210		
	Chert, ferruginous pelites, quartz pebble horizons	
1010		
	Banded ironstone with interbedded ferruginous phyllites	Unit reduced to discontinuous stringers near Mtshingwe Fault. Possible sulphide-rich horizons
900		
	Ferruginous pelite, chert	
870		
	Serpentinite	Localised pod
810		
	Ferruginous pelite with chert bands, quartz pebble horizons	Possible sulphide rich horizons
770		
	Mixed quartzite, arkose pebble conglomerate with chert clasts. Dominantly 5-10 cm bedded quartzite	Unit 300 m thick in north, 40 m thick in south just north of Mtshingwe Fault
550		
	Mixed quartz grits, siltstones, pelites minor cherts. Over 100 m of massive mafic and ultramafic intrusives	
	Vukwe Serpentinite	Unconformity or intrusive contact
0		

89

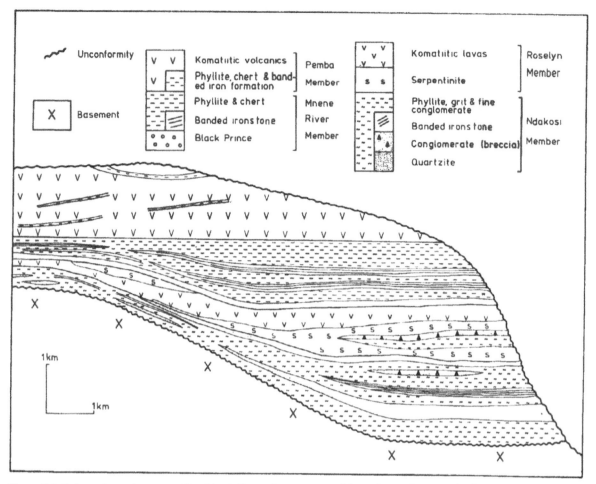

Figure 5.2. Schematic section across Brooklands Formation constructed from down-plunge projection along D_g1 axis with displacement along Mtshingwe fault restored.

locally pelites near igneous rocks contain pseudomorphs after andalusite, possibly formed in an aureole near an intrusion. It is possible that the igneous rocks are part of the Roselyn suite (see below).

The type section of the member is taken west of Vukwe Mountain where the unit is thickest (Table 5.1). There is very great lateral variation away from the type section. In the type section the base of the member rests against the Vukwe serpentinite; elsewhere it probably originally lay on ancient (3500 Ma) gneissic crust.

The Ndakosi Member shows great lateral variation, thinning rapidly southwards. Although part of this thinning may reflect variation in tectonic strain, the facies changes (especially in the conglomerates, which are much more common in the north) show that the thickness variation is an original sedimentary feature. The regional variation is demonstrated clearly by the lower quartzite unit, which ranges from 300 m thick in the north of the area to 40 m thick just north of the Mtshingwe Fault. The Violet Whoopee (V.W.) quartzite breccia at 1570 m in the type section thins rapidly laterally and has a strike length of 1100 m. Within the horizon, clast size also decreases rapidly laterally. In the north of the area is a distinctive conglomerate containing 2-5 cm long, well-rounded pebbles set in a chlorite-muscovite-quartz matrix. This may be a felsic volcanic horizon, possibly a lapilli tuff. Locally this unit reaches 400 m thick, elsewhere it outcrops as thin (5 m) bands.

The Violet Whoopee and Asci beds (Table 5.1) are discussed in detail later below.

South of the Mtshingwe Fault the succession is thinner and dominated by chlorite-rich pelites, silts and grits. In the south of the area a distinctive breccia of sub-angular 3-20 cm long vein quartz clasts set in a clinozoisite-quartz rich matrix occurs as a basal lithology 10 m above gneiss outcrops. This may be a basal breccia on 3.5 Ga gneissic crust. It should be noted however that the quartz clasts are highly strained and the breccia may be of tectonic origin. Not far above this breccia 20 cm-thick trough cross-bedding is well preserved in an isolated sandstone outcrop suggesting shallow water conditions above the possible unconformity. Near the top of the Ndakosi Member are pure massive quartzites, and associated sandstones with flaser bedding, ripple-drift lamination and 5-20 cm thick trough cross-sets. All indicators young west.

5.2.2 *Roselyn Member*

The Roselyn Member consists mostly of greenschist-grade meta-igneous rocks, which are only briefly described here. The type section is in the Ngezi River (Table 5.2). The komatiitic lavas show pillow structures and have MgO around 20% (Chapter 8). They are associated with massive ultramafic rocks, typically serpentinised with serpentinite-chlorite-talc-dolomite assemblages. These rocks may be of intrusive origin and are mostly absent in the south of the area.

5.2.3 *Mnene River Member*

The Mnene River Member consists of basal grit, a distinctive conglomerate horizon (the Black Prince Conglomerate bed), siltstones, cherts and ironstones. The type section (Table 5.3) is west from the Ngezi River. The Black Prince Conglomerate bed is remarkable for its great lateral continuity. It is continuous from the south of Brooklands Farm to where it is truncated by the Manjeri Formation in the northwest, a total distance of 8 km. In this distance it shows little variation in either thickness or rock-type and, what thickness variation there is, is probably partially controlled by tectonic rather than sedimentary variation.

North of the type section the Mnene River Member appears both to thicken and to increase in sedimentary complexity. From south to north the few metres of basal grits change to silts overlain by 30 m of quartzite with 1 mm diameter subrounded quartz grains and some coarser grit bands (5 mm diam. grains). Locally the quartzite is interbedded with cherts. It is overlain by pelites and silts and then the Black Prince Conglomerate, which is in turn overlain by 1 km of siltstones and banded ironstone. Although the

Table 5.2. Roselyn Member.

Height in section (m)	Roselyn Member type section	Regional variation and comments
450		
	Grits of Mnene River Member. Carbonated ultramafic rock	Ultramafic rock limited in lateral extent
350		
	Komatiitic volcanic rocks (ca. 20 % MgO) pillowed and thin lava flows Quartzite of Ndakoski Member	Massive serpentinites, mainly north of type section, up to 750 m thick. Probably derived from dunite intrusive bodies
0		

Table 5.3. Mnene River Member.

Height in section (m)	Mnene River Member type section	Regional variation and comments
1200		
	Banded ironstone, some interbedded siltstone and pelite	
900		
	Interbedded siltstones, chert and thin horizons of banded ironstone	In north, banded ironstone and interbedded siltstones and pelites increase to over 1000 m in thickness
700		
	Banded ironstone	
300		
	Finely bedded siltstones and cherts, passes up into overlying banded ironstones	
	Grits	
100		
	Black Prince Conglomerate	
25		
	Grits	30 m thick quartzite with thin chert-pebble conglomerates in north
	Top of Roselyn Member	
0		

Table 5.4. Pemba Member.

Height in section (m)	Pemba Member type section	Regional variation and comments
	Reliance Formation komatiites	Unconformity or disconformity.
1650		Difficult or impossible to distinguish between Manjeri and Pemba
	Banded ironstone and pelites of Manjeri Formation	Member sediments. Reliance Formation volcanic rocks form a dis-
	Pelite, chert and banded ironstone predominant	tinctive unit above the sediments
1450		
	Thin clinopyroxene spinifex textured flows capped by thin cherts	Thin banded ironstone, pelite and chert horizons throughout member
1250		
	Pillowed komatiitic basalt volcanic rocks, tholeiitic basalts	Mostly massive structureless mafic mafic lavas with occasional pil-
850	East Dyke	lowed outcrops and large quartz-filled lava withdrawal channels
700	Structureless mafic rocks possibly lava flows, thin pelites and cherts	
0	Mnene River sedimentary rocks	

main ironstone horizons can be traced several kilometres, individual layers are often impersistent. It is probable that some bodies mapped as 'chert/ironstone' may be silicified zones in pelitic sediments. Also, as exposure is incomplete, it is impossible to distinguish tectonic breaks from sedimentary discontinuities. In places, two separate ironstone horizons merge, cutting out pelitic intervals. The number of mappable ironstone horizons increases to the north, suggesting that part, at least, of the increase in thickness of the member as a whole, is a primary sedimentary feature.

South of the type section the member is approximately 700 m thick and the proportion of banded ironstone is very much less.

5.2.4 *Pemba Member*

This member consists of a second unit of mafic volcanic rocks, associated with minor cherts, banded ironstones and pelites. The type section (Table 5.4) is taken from the major bend in the Mnene River up a small stream to the south-west. At the base are pelites and banded ironstones, which pass up into mafic rocks, mostly massive but locally pillowed and near the top showing spinifex textures after clinopyroxene. Overlying the member, above the unconformity, are Ngezi Group sediments. Away from the type locality some ultramafic material is present.

5.3 SPECIFIC FACIES TYPES

Two questions may be posed in an analysis of the Brooklands Formation: 1. Was the formation deposited on a pre-existing ancient basement? 2. Does the sedimentology of the formation give any clue as to the tectonic processes controlling deposition?

Four distinct facies associations were chosen for study in detail: A breccia facies; a conglomerate facies, a siltstone/shale facies; and an ironstone facies. Not all the sediments of the formation fall into these categories and detailed study was necessarily confined to the few best outcrops of chosen rocks.

5.4 THE BRECCIA FACIES

Several distinctive breccia units occur in the upper part of the Ndakosi Member (Fig. 5.3). These breccias show rapid lateral facies changes and locally contain very large angular clasts set in a chlorite-grit matrix. We have studied two units in some detail.

5.4.1 *Violet Whoopee breccia*

This unit is named after the Violet and Whoopee claims (now lost) registered on Brooklands Farm. Exposure is poor except in one donga-section (a donga is a local name for an incised stream-bed) at GR 004310. The breccia is difficult to trace laterally. It probably ranges from 50 m thick (maximum) to 10-20 m at the best outcrop, tailing off to a thin band in the poor exposure. The unit has a mapped length of 1 km, giving a very rough thickness to length ratio of between 10:1 and 40:1 in the plane of the outcrop. Internally, the unit appears to be massive.

The country rock is typically a chloritic phyllite, semi-pelite or psammite, containing occasional quartzite bands, quartz lenses, veins and pebble beds. The bulk of the country rock was probably originally an arenite or siltstone, with a major component of lithic

Figure 5.3. Asci breccia, Ndakosi Member. Note angularity of folded gneiss clast.

Figure 5.4. Violet Whoopee (V.W.) breccia. a) Clasts of quartzite in chloritic (mafic) matrix. b) Clasts up to 1.5 m long.

fragments as well as quartz, derived from a varied granite-gneiss-greenstone terrain. The country rock appears to be generally similar in appearance to the matrix of the breccia bed.

Clasts in the breccia show little or no rounding, tend to be tabular, are often large, and are randomly distributed in the matrix (Fig. 5.4). They are of quartzite and were possibly only partly indurated prior to erosion and redeposition as some clasts appear to have been folded during sedimentation. Some tabular clasts appear to be broken up, but the fragments appear to have still maintained their relative positions, although now separated from each other (at least in the plane of outcrop) by the matrix. Clast lengths range up to 1 m or more (Figs 5.4a, b). The largest clasts contain the majority of the clast material (Fig. 5.5); in volumetric terms the numerous small clasts are relatively insignificant, although their importance may have been slightly underestimated as they are difficult to identify.

Two measured lines across the surface of the outcrop (Fig. 5.7) show the angularity of the clasts, their size and packing, and their orientation. In general, orientation (Fig. 5.6) appears to be random, although large clasts appear to have 'domains' of influence in which the orientation of minor clasts is controlled by the major clasts. Structural complexity hinders any further interpretation of orientation.

The rock is very strongly bimodal, with an arenaceous matrix and clasts which average 1250-2000 cm^2 in the plane of outcrop. In outcrop 50% (very roughly) of the surface area is clast, the rest matrix. The rock appears to be matrix-supported, especially given that most clasts are tabular (and would thus pack well). The matrix consists of mainly lithic detritus, quite different from the clasts. This contrast, and the matrix-supported nature of the rock make it unlikely that the rock is a tectonic breccia.

5.4.2 *Asci breccia*

The Asci breccia forms part of the upper Ndakosi Member (GR 975296). The Asci breccia is broadly similar to the V.W. breccia, but clasts are typically somewhat smaller in the outcrops studied and a much greater lithological variety is present. In Figure 5.3 a fragment of foliated gneiss can clearly be distinguished. Gneiss and other fragments often have very straight edges, as if they had been fractured from joint and foliation surfaces and transported without rounding.

The matrix forms 50% of the surface area of the outcrop (visual estimate) and the rock, in part at least, is matrix supported and very strongly bimodal. The matrix consists of abundant crystals derived both from a probable mafic source and from a possible gneiss or granulite source. Metamorphic rock fragments generally contain highly strained quartz or consist of aggregates of sutured quartz. Feldspar is common in the matrix. The igneous detrital material appears undeformed in contrast to metamorphic clasts. Possible uralitised spinifex pyroxene crystals suggest a komatiitic source terrain.

5.4.3 *Interpretation*

There are two major processes which could have produced these breccias – glacial or peri-glacial sedimentation, or mass-flow and catastrophic processes. How does one distinguish between these alternatives? To review the available data:

a) The breccias are massive. They appear to show no sedimentary structure at all, such as bedding planes or grading, in the present outcrops.

b) Clasts range up to 1.5 m across: The breccias are very coarse locally.

c) Clasts are almost always angular.

d) In the Asci breccia clasts come from a variety of source rocks.

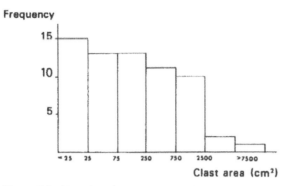

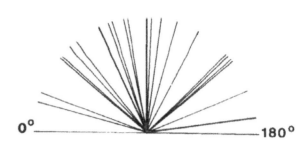

Figure 5.5. Clast size distribution in V.W. breccia. Surface area measured in cm.^{2} Frequency is number of clasts of given surface area in sampled area.

Figure 5.6. Orientation of clasts in V.W. breccia. Note lack of preferred orientation.

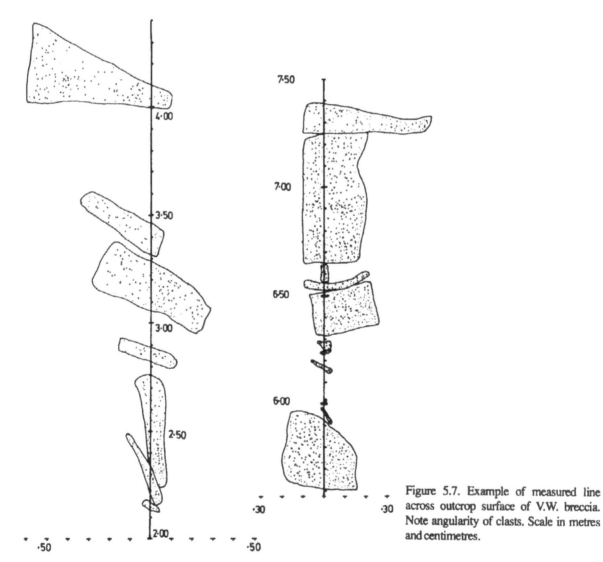

Figure 5.7. Example of measured line across outcrop surface of V.W. breccia. Note angularity of clasts. Scale in metres and centimetres.

e) In the V.W. breccia the orientation of small clasts can be controlled by the larger clasts, indicating some slight shearing during transport. However, some clasts have become disaggregated, yet fragments seem to retain their original positions relative to each other. Some may have been folded at some stage during transport.

f) The breccias appear to be matrix-supported, at least in part.

We consider that these characteristics, especially the last, imply that the formation was not in an immediate glacial or peri-glacial setting (e.g. as a moraine or outwash gravel). Kurtz & Anderson (1979) have considered the problems of distinguishing between orthotills, paratills and debris flow deposits in the Antarctic. They pointed out that it is very difficult indeed to distinguish orthotills (the marine tills deposited by a grounding ice sheet on the continental shelf) from debris flow deposits: However, orthotills do often have a crude pebble fabric parallel to bedding planes. Paratills (sediments resulting from the combination of normal marine sedimentation and deposition of clasts from floating ice) generally have a crude to good pebble fabric. It is unlikely that either of the V.W. or Asci breccias is a paratill. The evidence for some minor control of small clast orientation by large clasts would imply that shear took place between clasts during deposition: Because of this and because of the general absence of a pebble fabric even with these very tabular clasts we consider that the deposit was produced by a mass flow or slumps, not as an orthotill. The lack of scattered boulders in the country strata also argues against glacial conditions; some paratills would be expected if the breccia is an orthotill. Thus the V.W. and Asci breccias were probably emplaced as debris flows, slumps or avalanches.

Johnson (1970) has given detailed descriptions of debris flow deposits. Transport may be for considerable distances and only stops if the slurry dries out (subaerially), meets a slope inversion or becomes too dispersed to continue moving. In the Brooklands Formation breccias relative motion between clasts was probably very restricted, since disaggregated clasts are found with fragments still lying in what appear to be their original positions relative to one another, yet some shear must have taken place to allow large clasts to control the orientation of small clasts. Very few clast-clast collisions could have taken place during transport.

5.4.4 *Erosional and depositional settings*

A variety of mass flow or slump processes can produce the type of deposit observed in the Brooklands Formation. These processes include large rock avalanches, debris flows, talus cones, rock slumps and slides (Yarnold & Lombard 1989; Prior & Doyle 1985). Large rock avalanches tend to produce thick sheet or tongue-like deposits of clast supported sediment; debris flow deposits are usually less than 5 m thick and matrix supported; talus cones are clast supported with little primary matrix, and slumps and slides produce large coherent blocks of rock with little matrix. Most likely, the bodies seen in the Brooklands Formation are debris flow deposits, though some may be avalanche or rock-fall deposits. This interpretation is based on the thickness and shape of the bodies, the role of the matrix and the local stratigraphy (Nardin et al. 1975).

Williams & Guy (1973) and Campbell (1975) described the depositional consequences of major flooding in Virginia and California. Avalanches described by Williams and Guy ranged in size from 60-250 m long, 6.25 m wide and up to 5 m deep, with average gradients ranging from 16-39x. Similar features are widespread in the Alps. Avalanche volumes ranged from 525-670 m^3 in the Virginia study. Deposition occurred in five main types of sites (in order of decreasing clast size): a) Debris avalanche deposits at the base of hill-slopes, b) mountain channel deposits, c) alluvial fans, d) deltas where the flow was blocked by an obstruction, and e) vertical accretion deposits on flood plains.

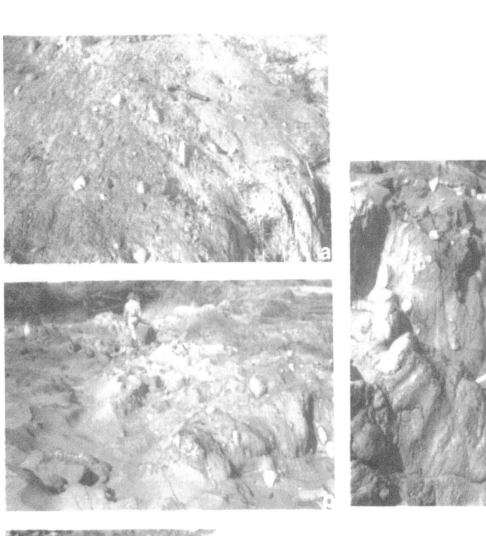

Figure 5.8.a) Lower Ngezi outcrop of Black Prince Conglomerate, Ngezi River. Stratigraphic facing to left. b) Upper Ngezi outcrop of Black Prince Conglomerate, Ngezi River. Note intrusion cross-cutting conglomerate (on left). c) Detail of outcrop in a. Note intraclasts. d) Example of clast extracted from Black Prince Conglomerate. Note that elongation reflects tectonic strain. Mnene River outcrop.

In the first three categories clasts up to 1 m long were observed. Deposits were mostly unsorted, though some showed lenses of pebbles in an imbricate pattern. From photographs in the two works cited it is clear that many clasts were tabular in shape and unrounded. Alluvial fan deposits showed few sedimentary features. Grain sizes ranged from 2-1700 mm across. The size of the largest clasts gradually decreased downfan. Sorting was poor in fans and they consistently had polymodal size distributions with a scarcity of particles in the range 4-64 mm diameter. This distribution is similar to that visually estimated in the V.W. breccia. Both works cited stressed the importance of heavy rainfall in triggering landslips, though Johnson (1970) described flows triggered by flooding from melting snow.

Less is known about submarine debris flows, but they too can probably move very large clasts, without rounding them. Submarine rock-fall and slump deposits have been observed by Prior & Doyle (1985).

The reason for supposing that the deposition of the breccias was probably subaqueous is that the country rock is of probable subaqueous origin and there is no evidence to suggest any brief emersion. Thus one possible scenario for the formation of the breccias is that they were produced by debris flows originating on land, possibly triggered by slumping on fault scarps after earthquakes or heavy rainfall, with debris moving by mass flow to a nearby lake or sea, where the deposit was preserved. It should be noted these must have been rare events, as the number of breccia beds is only a minute proportion of the total number of beds in the country rock.

In the V.W. and Asci breccias, however, the extreme angularity and size of the clasts implies that if the material was deposited subaqueously, the clasts may also have been removed from the parent rock subaqueously. It is unlikely that they could have been part of a debris flow carrying detritus which had previously been transported from the land by fluvial processes. However in most debris flows only surface weathered material is stripped off – the flows do not cut deeply into the parent rock. Whether Archaean subaqueous weathering penetrated deeply enough to break off angular fragments of rock without chemically degrading them is unknown, but it is possible that a subaerial environment would have been more capable of producing the large clasts which appear to have been mechanically weathered.

This interpretation is consistent with the strong evidence that the source of debris was very close by. Williams and Guy proposed empirical relationships between size of clast and distance of transport. These relationships when applied to the breccias suggest that the source was perhaps as close as a few kilometres away.

The deposits may have had a wholly subaqueous origin, as slumps and rockfalls off submarine scarps (Prior & Doyle 1985). The degree of facies variation in the sedimentary sequence in the Brooklands Formation is evidence in support of the notion that marked subaqueous topography existed. An alternative possibility, raised by H. Burgisser (pers. comm.) is that these may be breccias transported from a cliffed shoreline (Lawson 1976). The breccias are small, local, and not very common (<1% of total sequence) in a sequence originally mainly consisting of pebbly sandstone, pebbly siltstone and pelite. Chaotic shoreline breccias would provide a suitable occasional source for material transported by debris flow to a site of deposition in slightly deeper water.

The source terrain must have been very variable geologically. Most probably it consisted of an ancient granite-gneiss terrain, possibly with greenstone relics. However, the igneous and quartzite clasts were probably derived from a cover sequence as they appear less deformed. Quartzite clasts in the V.W. breccia may not have been fully indurated when eroded, a further point in favour of a subaqueous origin, as rockfall and slumps. Thus much of the sedimentary material may have been very locally derived.

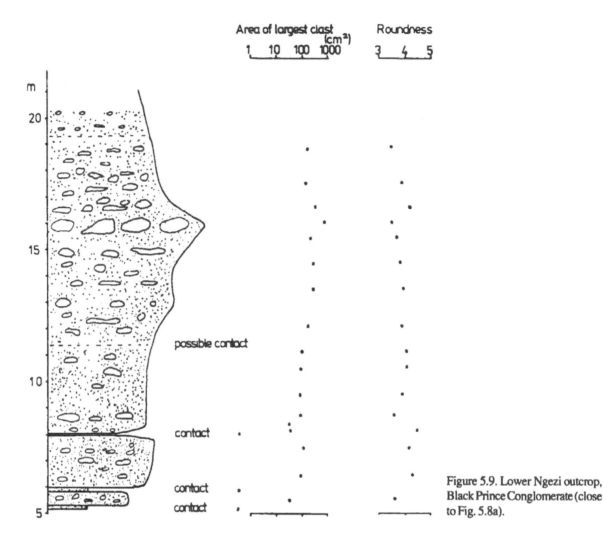

Figure 5.9. Lower Ngezi outcrop, Black Prince Conglomerate (close to Fig. 5.8a).

5.5 THE CONGLOMERATE FACIES

5.5.1 *The Black Prince Conglomerate bed*

The Black Prince Conglomerate (named after the lost Black Prince claims on Brooklands farm) forms part of the type section of the Mnene River Member of the Brooklands Formation. The conglomerate is very continuous laterally. It can be traced over a distance of 8 km, and has a thickness of 20-50 m. The thickness does not appear to vary greatly along strike, although exposure is poor in many places. It is not clear how much of the thickness variation in the range of 20-50 m can be ascribed to original sedimentary variations, and how much is a result of tectonic strain. The considerable strain that the rock has suffered (see Chapter 3) makes the identification of original sedimentary fabrics difficult and makes a study of clast orientation unprofitable. However, it is unlikely that a major variation in original thickness (such as the valley of an ancient submarine canyon) could have been removed by later strain – we thus conclude that no such original feature is exposed.

In general the Black Prince Conglomerate is underlain by arkose grits and phyllites of the Mnene River Member. Above the conglomerate are arenites, passing rapidly into

100

siltstones and shales, followed by ironstone (see below). The conglomerate has been studied in some detail at several outcrops. Two outcrops in the Ngezi are particularly well exposed. The outcrops are shown in Figure 5.8. We shall describe each outcrop separately and attempt to synthesise the data to construct a model of the depositional environment of the unit. In the following discussion the term 'grit' (Old English greot – sand or gravel) is used for a rock consisting of angular particles approximately 2-4 mm diameter.

5.5.2 *Lower Ngezi outcrop*

The thickness of the conglomerate sequence in this outcrop is difficult to determine precisely – at least 20 m are well exposed and the total thickness is probably about 50 m. The outcrop is dominated by a series of layers of coarse conglomerate. These layers are sometimes separated by thin sandy beds and lenses. Few erosional features are visible in outcrop. One channel cut into the conglomerate is discussed below.

The rarity of erosional features makes the identification of individual conglomerate deposition events difficult. The following criteria were adopted in recognising separate events:

a) The presence of an erosional feature, or, if absent,

b) abrupt changes in lithology – for instance the change from a coarse boulder conglomerate to a cross-bedded arenite, or, if absent,

c) abrupt changes in clast size – for example the change from a cobble conglomerate to a boulder conglomerate over an interval of, say, 1-3 cm, if this change is laterally continuous.

Criterion (c) is most difficult to apply. In general the conglomerate clast size changes only slowly through a recognised unit. However, it is possible that sudden changes in depositional process, e.g. from mass-flow transport and deposition to current deposition, may have taken place within a single event. In the discussion which follows the three criteria were applied conservatively, for example in the study of grading, great care was taken to establish that only positively identified depositional units were used. The units in which grading was studied may have been thicker than has been assumed in the following discussion (which would further support the identification of inverse grading), but only the positively identified portions of the units were used to identify grading patterns.

Sedimentary column

Figure 5.9 shows a column through the outcrop. The traverse was chosen to sample a 'typical' section of the outcrop and does not include the largest boulder seen.

At the base of the column siltstone passes to 5 m of arenite. This arenite is finely laminated where exposure is good, with laminae 0.1-1 cm thick and probable cross-set units 10 cm thick. At 5.2 m is a thin grit band, overlain by a 50 cm unit of pebble conglomerate/pebbly mudstone, followed by a further grit band. From the top of this grit to 22 m in the sequence the rock is thought to constitute a single conglomerate unit and there are no major breaks in the sequence. However, application of criterion (c) above suggests very weakly that distinct conglomerate units may have been superimposed without erosion at levels 11.4 and 19.3 m in the sequence. The change in fabric is not very marked at these levels and it is not possible to decide whether the full unit stretches from 8.1 to 22 m (Fig. 5.9) or whether there are three units. Thus the sequence from 11.4 and 19.3 m is regarded as one positively identified unit, with the possibility that the unit stretches from 8.1 to 22 m.

The largest clasts are found around 16 m in the sequence. Thus in either case, whether 8.1 or 11.4 m is taken as the base of the sequence, inverse grading does exist in the lower portion of the unit.

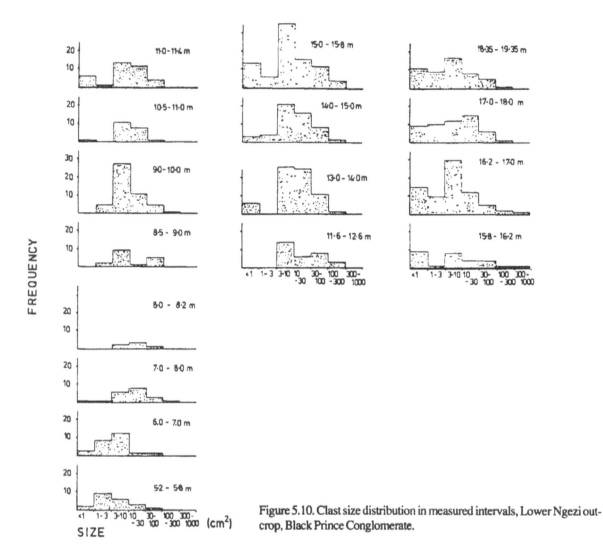

Figure 5.10. Clast size distribution in measured intervals, Lower Ngezi outcrop, Black Prince Conglomerate.

Above 16 m (Fig. 5.9) the conglomerate becomes progressively finer and clasts become more angular, and sometimes broken. Above 22 m exposure is poor.

Sediment fabric
Orientation. The rock has undergone considerable tectonic strain, which has almost certainly rotated the clasts. In the absence of a detailed study of strain markers no conclusions may be drawn about clast orientation before deformation.

Clast composition
Five different classes of clast were recognized in the field, and a 1m square portion of the outcrop (offset from 17 to 18 m in Fig. 5.9) was studied in detail to determine clast composition. Of the 126 clasts studied, half were of mafic igneous rock or chloritic schist, a fifth quartzitic, a fifth mapped in the field as arenaceous, and the remainder were coarse quartz-rich rocks (possibly including intraclasts of redeposited conglomerate and plutonic igneous material) or vein quartz detritus. Rare ironstone clasts were seen elsewhere in the outcrop. The clasts of quartz and quartzite were readily detected as they weather out of the matrix, but clasts of other materials generally weathered flush with the matrix and were difficult to distinguish from it.

102

In thin section the clasts were dominantly quartzite and komatiitic basalt. Komatiitic clasts varied considerably, from altered and carbonated material in which actinolite had replaced clinopyroxene laths while relict textures remained, to fragments with plumose clinopyroxene bundles, to fragments with clinopyroxene microlites and laths set in isotropic altered glass.

Because of the difficulty of identifying non-quartzitic clasts in outcrop, it was not possible to study the variation in clast composition across the whole outcrop. There was no visual evidence of significant variation.

In thin section the matrix of the rock was dominated by the breakdown products of igneous rocks, single pyroxene crystals and quartz and feldspar crystals. Some quartz was strained, indicating a previous metamorphic history.

Roundness

Clast roundness is typically 3 to 5 (Fig. 5.9) estimated according to the Powers scale (as illustrated in Blatt et al. 1972), using a large sample of clasts in each segment of the measured column, with the results averaged in the knowledge that precision is poor. Only quartz and quartzite clasts were used in the roundness study as mafic clasts were difficult to identify and were deformed. In general, mafic clasts appeared to be much more elongate, perhaps as a result of deformation, as mafic clasts appeared to be moulded around quartzitic clasts. Within the limits of precision (perhaps half a class) there is no significant variation in roundness except near the top of the section, and there was no evidence to suggest that mafic clasts showed any significant variation either. The top of the succession contains rather more angular clasts, and the 15.8-16.2 m segment with the largest clasts was also perhaps slightly less well rounded than average.

Grading

To investigate grading the traverse was divided into the same segments as in the roundness study. Where obvious sedimentary breaks occurred, these were taken as segment boundaries; elsewhere boundaries were arbitrary. In each segment the surface dimensions of all boulders projecting from the matrix (i.e. quartz and quartzite clasts predominantly) were measured. The surface area of each clast was then calculated, assuming an elliptical cross-section in the plane of outcrop. For each segment the total area of clasts was calculated and divided into size groupings.

The results of the measurements are shown in Figures 5.10, 5.11, 5.12. Figure 5.10 shows histograms of clast abundance in each of the segments. The size intervals are logarithmic. The inverse grading in the interval 11-16 m can be clearly seen, and the absence of large clasts above this. If the unit begins at 8.4 m the inverse grading stretches over about 7.6 m, followed by at least 2.3 (more probably 7) metres of normal grading. Figure 5.11 shows the grading in more detail, giving total area of clasts in each of a set of size groups.

Figure 5.12a shows the normalised distribution of clast size (without matrix) for the set of arbitrarily chosen size groups. The horizontal axis is the area of each size group; the vertical axis is the total area of each size group, normalised to 1 m^2 and by the average clast area of each group to make the data comparable. In this sequence 11.4-19.35 m all segments show very similar distribution patterns, strongly suggesting that they were laid down in the same process. The inverse grading may be seen by examining the larger size groups. It is interesting to note that the highest volumes of clast occur in segments below the segments which contain the largest clasts. The topmost segments contain much smaller proportions of clasts. The implication is that in the upper segments the flow was less able to sustain a high clast/matrix proportion.

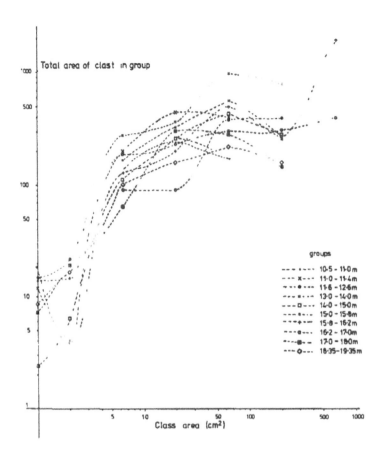

Figure 5.11. Plot of total clast area in unit area of outcrop in size groups for measured stratigraphic intervals, Lower Ngezi outcrop, Black Prince Conglomerate. For each interval clasts were divided into a set of size classes (1-5, 5-10, 10-50, 50-100, 100-500 cm^2) and total area in each class was determined. Note predominance in area (and hence volume) of the larger clasts. The plots show intervals of normal and inverse grading.

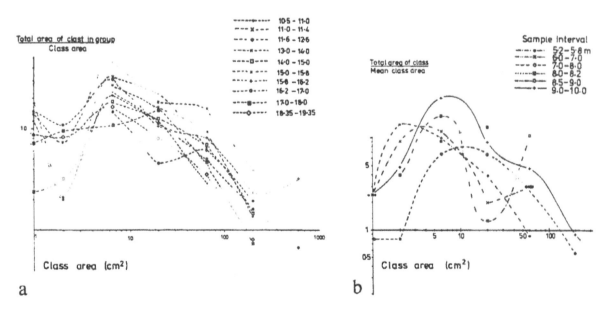

a   b

Figure 5.12. a) Clast size distribution, Lower Ngezi outcrop, Black Prince Conglomerate. Same data as Figure 5.11, but normalised to average clast area. Histogram of normalised total area of clast in chosen size groups for vertical segments of outcrop. Average class area is average of clasts falling in particular size grouping. Note grading. b) Same plot as a : Lower part of unit.

Figure 5.12b shows the same plots for other segments. The general similarity of the patterns for segments above 8.1 m to the segments in Figure 5.12a suggests that they all constitute the same single flow unit (although one point in the 8.5-9.0 m segment is not consistent with this). Lower segments known to come from different flows have different patterns. These diagrams were plotted from measurements of clasts which weathered out from the matrix – dominantly quartz and quartzite. Since 50% of the clasts are mafic, the figures may not give an accurate idea of clast distribution within each segment, although as a general observation the sizes and proportions of mafic clasts closely parallel the quartzitic clasts. To test this observation a segment at the 17-18 m level was studied in great detail (Fig. 5.14). This segment was offset a few metres from the 17-18 m segments in the column previously discussed, on the same stratigraphic horizon. Figure 5.13 shows the results. In the 1 m square offset segment all clasts were carefully identified. Comparison of Figures 5.10 and 5.11 with 5.13 shows that in the measured column the significance of small clasts was seriously underestimated, but the abundance pattern of the major (volumetrically important) clasts shows no significant difference in pattern from the data from the main measured column, indicating that the conclusions drawn from the main column are valid. Since the larger clasts dominate the total clast content overwhelmingly, the underestimation of small clasts is of relatively minor importance in the measurement of total clast area.

Packing proximity
A series of traverses was made normal and parallel to strike to measure packing proximity. Traverses normal to strike gave average proximities around 10%, while traverses parallel to strike gave a range from 0-28%. These traverses were made in the intensively studied area of Figure 5.9.

It should be noted that tectonic strain may have had a significant effect on packing proximity, especially since strain was probably very inhomogeneously distributed. In general, all that can be said about packing is that it was probably originally in the range 0-33%. Despite the apparently low packing proximity it is probable that the conglomerate is clast supported.

Erosional features
As stated above, most of the lower Ngezi outcrop of the Black Prince Conglomerate shows no erosional features at all. However, one portion of the outcrop near the base of the succession (below the main flow unit discussed above) shows a channel cut into relatively fine conglomerate (Fig. 5.15). The channel contains clasts up to 10 cm across, implying currents perhaps of the order of 5 m/s (Hjulstrom/Sundberg diagram).

All material of the channel-fill appears to be very locally derived, most probably from the top of the underlying flows.

Sandstone lenses
Some thin (ca. 1-10 cm thick) but persistent sandstone lenses occur (Fig. 5.16). They can be traced for 20 m or more across outcrop, separating the flow units, and implying that the agent that deposited the flow units had a low basal erosional capacity. Laminations in the lenses indicate that they were laid down by current processes. The lenses vary little in thickness along their length, implying flat topography. The thinness and relative scarcity of the lenses implies that conglomerate flows followed each other very rapidly. Some flows may have been erosional at their bases, removing lenses, but the very existence of the lenses implies little erosion. The lenses appear to represent episodes of current deposition between conglomerate flows.

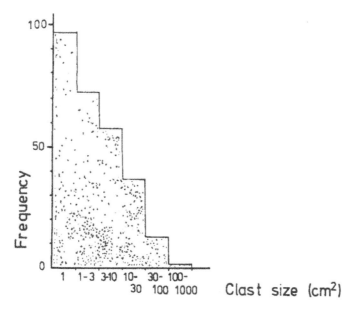

Clast size (cm²)

Figure 5.13. Clast size distribution in one very carefully studied segment, at 17-18 m in column, Lower Ngezi outcrop, Black Prince Conglomerate. Contrast with Figure 5.10: 17-18 m.

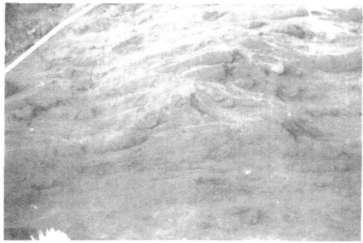

Figure 5.14. Area studied in great detail, Lower Ngezi outcrop, Black Prince Conglomerate.

Figure 5.15. Breccia channel cut into flow unit, Lower Ngezi outcrop, Black Prince Conglomerate.

5.5.3 *Upper Ngezi River outcrop*

The general aspect of the conglomerate in this outcrop is very similar to that in the lower outcrop. Figure 5.8b shows the outcrop and Figure 5.17 is a section through it. A notable feature is a 5 m thick mafic body. The conglomerate outcrop is about 40 m thick, but is partly hidden in thick bush.

No obvious sedimentary breaks can be distinguished in the 10 m of sequence studied in detail. Figure 5.18 shows clast size variation. No channelling or other erosive features are present. The variation in clast size indicates that there may be two distinct flow units in the sequence. The unit from the top of the mafic sill to 9 m shows probable inverse grading. Roundness is generally very high throughout the outcrop, with no significant variation. Packing is similar to the Lower Ngezi outcrop. The largest single clast measured $60 \times 30 \times 47$ cm, another $85 \times 53 \times 18$ cm.

One small portion of the outcrop was studied exhaustively, as in the Lower Ngezi outcrop. Most of the clasts in this segment were mafic or chloritic schist (together these were 62% of the total sample), and most of the remainder were quartz-rich (36%). Some sandstones and some vein quartz clasts were noted. The composition of the clasts suggests a slightly different source compared to the Lower Ngezi outcrop, or local variation in weathering or transport processes during sedimentation.

5.5.4 *Other outcrops studied*

Four other outcrops were studied. All were broadly similar to the two Ngezi outcrops. Outcrop A, at GR 956300, is poorly exposed, the thickness of the body appears to be thinner than in the Ngezi outcrops (47 m), and clasts appear to be smaller, but this may be a result of poor outcrop. Quartzite clasts are well rounded, and range up to 30×15 cm in outcrop, averaging 5×3 cm.

At outcrop B (GR 961294), the section is the thickest measured (140 m). This locality lies in the hinge region of D_g3 folds and it is probable that much or all of the thickness variation is due to strain. At this outcrop the full sequence through the conglomerate may be seen. At the base of the conglomerate is a pebbly arenite/grit containing clasts around 1 cm long. Many clasts are poorly rounded to angular, often elongate. Clasts of sandstone are common; they range up to $6 \times 4 \times 0.5$ cm in size.

The conglomerate proper contains notable quartzite clasts up to $30 \times 35 \times 20$ cm size, but averaging about 5×3 cm in outcrop. It is difficult to estimate the relative importance of other clast materials as the outcrop is rather weathered. The rock appears to be massive and lithologically very similar to the other outcrops. The conglomerate grades up into grit and pebbly mudstone, often with bands of pebble-rich rock set in an arenaceous matrix, interspersed with bands of ferruginous phyllite containing occasional randomly distributed small pebbles typically 2 cm across (rarely 5 cm). Pebbles are made of quartz and possibly other rock types (weathering obscures identification).

Mnene River outcrop C (GR 968257), is some distance from the others studied. At the base of the outcrop is an actinolite-schist, overlying recognisable mafic igneous rock. The schist passes into a 3 m thick conglomerate unit, thought to represent a single flow event. This unit is inversely graded at its base. The largest clasts are in the centre of the unit; above this, grain size decreases upwards. Above this unit is a grit band, which passes into a 17 m thick massive conglomerate with clasts up to $40 \times 20 \times 30$ cm. Packing proximity is around 20% and roundness is very high, averaging 4.4. At 17 m the sequence passes to pebbly mudstone. Nearby to the main outcrop in a small stream 20 m of conglomerate are exposed, followed by a 50 m gap in outcrop. The thickness of the unit is thus in the range 17 m to at least 20 m.

Figure 5.16. Sandstone band (under tape), Lower Ngezi outcrop, Black Prince Conglomerate.

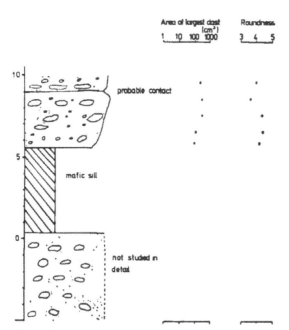

Figure 5.17. Upper Ngezi outcrop, Black Prince Conglomerate (see also Fig. 5.8b). Vertical scale in metres.

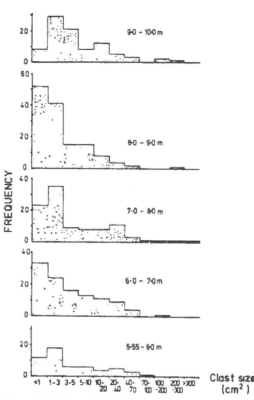

Figure 5.18. Clast size variations, Upper Ngezi outcrop, Black Prince Conglomerate.

5.5.5 *Interpretation*

Six separate lines of evidence exist in the Black Prince Conglomerate:

    a) Clast composition and matrix composition.

    b) Clast rounding.

    c) Clast size and grading.

    d) Packing; clast support of the rock.

    e) The lack of erosional features.

    f) The lateral persistence of the conglomerate unit as a whole.

Source terrain

Some of the clasts appear to have been derived from a basalt-komatiite terrain: detrital matrix comes from a similar terrain and also possibly come from an ancient gneiss/granite terrain. It is reasonable to suppose that this ancient gneiss/granite terrain, and probably also the komatiitic terrain, were broadly the same as the source terrain of the V.W. and Asci breccias, either the 2900 Ma terrain or the 3500 Ma terrain or both. In Archaean granite/gneisses, surface weathering probably produced abundant quartz-rich float (see description of basal Manjeri Formation unconformity) similar to the abundant quartz debris found today on the same terrain. This debris would have formed an important part of any sediment derived from erosion of the gneissic terrain. The source area also contained abundant mafic rocks. Some of the clast material is comparatively fresh and appears not to have been deformed prior to deposition; this igneous material may have been eroded from the immediately underlying Roselyn Member of the Brooklands Formation. Other clasts are more altered and could possibly represent an ancient greenstone terrain.

The clasts of sedimentary rock could also have been derived either by erosion of the lower parts of the Brooklands Formation itself, or from basement terrain. Further study of the metamorphic grade of the clasts may clarify this.

The matrix was derived both from mafic and granitic terrains, and igneous detritus forms an important component. It is possible that much of the matrix is derived from splitting of the clasts during transport, but the particle size of the quartz in the sandy matrix may also in large part reflect the grain size of the parent rock.

Transport processes

From the evidence presented it is clear that the final process of transport and deposition which laid down these rocks produced inverse grading. Several possible mechanisms of transport could produce this: Processes on beaches; glacial processes; transport in debris flows and transport in a grain flow. The highly rounded clasts in the conglomerate would imply that it is unlikely that this deposit is a till; it was probably laid down by a mass flow process, or on a beach.

A debris flow apparently produces little or no erosion of beds underlying it (Johnson 1970), and can also produce inverse to normal grading, with the largest boulders in the centre of the flow and smaller clasts above and below (Johnson 1970; Enos 1977). However, the dominantly clast-supported nature of the deposit, the lack of lateral thickness variation and the very even nature of the top and bottom depositional surfaces strongly suggests that grain flow, not debris flow, predominated as the method of transport.

A grain flow hypothesis would also account for the inverse grading (finer-coarser-finer). In grain flows the largest clasts occur in the middle of the flow or above. The relatively low packing, however, suggests that if grain flow were the main method of transport, much of the dispersion must have been by sand/clast collisions, although

Figure 5.19. Shale-siltstone facies, Ngezi River, upper outcrop. a) Note chert nodules. b) Laminated silt (white stars are reflections). c) Deformed laminations, shale-siltstone facies.

110

obviously clast/clast collisions must have been critical in maintaining the grain flow. If grain flow did take place, these collisions would have contributed to the high degree of roundness seen. However, in this case important lateral variations in clast/matrix and in sizes of the largest clasts would be expected, with fewer and smaller clasts in the more distal outcrops as compared to the more proximal. This has not been observed – all outcrops appear broadly similar, but exposure is often poor and deformation may have distorted the original packing, especially in the south of the formation. It is perhaps worth noting that Winn & Dott (1979), studying a late Cretaceous conglomerate in Chile, noted no lateral size grading in graded-to-massive conglomerate thought to have been deposited in fully turbulent flows. There are many parallels between the conglomerates described by Winn and Dott and the Black Prince Conglomerate. It is possible that the grain flows were set off by subaqueous landslips and debris flows, these then generating and being outraced by turbulent flows.

It is interesting to speculate on the roundness and composition of the clasts. A landslip or debris flow, such as the V.W. breccia may be, would produce extremely large angular boulders, probably with a limited range of compositions as the source terrain would be of limited extent. In contrast, transport of such material (perhaps originally stripped off the country rock by landslips or debris flows) by rivers or along shorelines would produce well-rounded and varied clasts. It is possible that final transport was by grain flow, triggered off by collapse of a coastal boulder deposit, or from a bedload brought down in flood. Final deposition was clearly subaqueous (see below for a description of surrounding strata). By this stage the clasts would be well-rounded indeed.

It is possible that all the structures seen were produced in a beach or near-beach environment. Such an interpretation would fit well with the planar nature of the Black Prince Member – it outcrops over a considerable distance, but varies very little in thickness, as would be expected in a uniformly transgressing beach. Tidal and perhaps fluvial transport would bring in a variety of clasts.

However, the regularity of the layering in the conglomerate bodies, and their individual lateral continuity argue against this interpretation. Furthermore, the close stratigraphic proximity of very distal facies makes a 'beach' hypothesis unlikely.

5.6 THE SHALE-SILTSTONE FACIES

Much of the Brooklands Formation is poorly exposed phyllite; generally this rock is badly weathered and original sedimentary fabrics are difficult to distinguish. In a few outcrops exposure is much better and information is available. The best exposure is in the Ngezi River, immediately above the upper exposure of the Black Prince Conglomerate. This exposure has been studied in detail.

The rock is well bedded green to red, typically ranging in grain size from clay to siltstone, and with frequent chert beds, lenses and stringers (Fig. 5.19). Occasional arenaceous beds also occur. The most notable features of the outcrop are the extreme persistence laterally of even the thinnest silt and clay beds, the often discontinuous nature of the chert, and the presence of slump horizons and early fracturing of the rock.

One set of beds, 30 cm thick, was chosen for a detailed study of the lateral variation and small-scale textures (Fig. 5.20). The beds were continuously exposed along a lateral distance of over 20 m, and the base of the 30 cm section was taken as the base of an easily traced continuous band. One column, Column D, was studied in considerable detail.

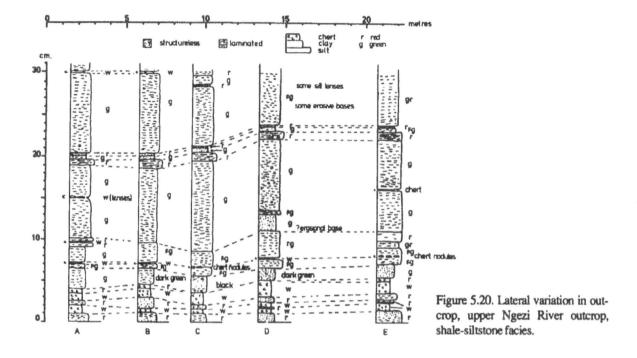

Figure 5.20. Lateral variation in outcrop, upper Ngezi River outcrop, shale-siltstone facies.

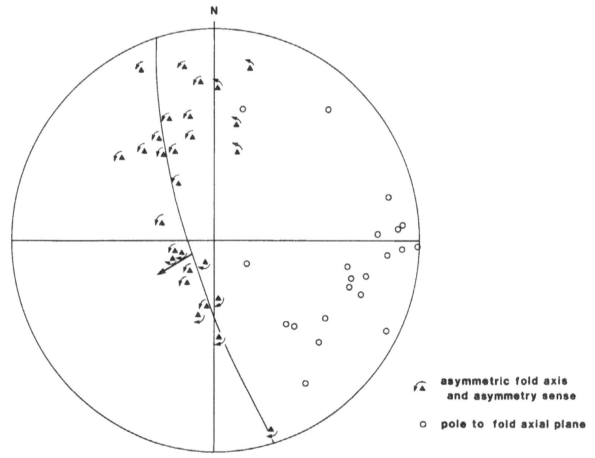

Figure 5.21. Fold hinge plunge orientations in shale-siltstone facies. Lower-hemisphere equal-area projection. Symmetry (overturning direction, see Hansen 1971) shown by arrow. Large arrow shows movement direction in plane of bedding of upper sheet.

5.6.1 *Fabric*

Lithology

Three rock types dominate the assemblage: Siltstone, shale and chert. Siltstones are typically (though not always) red, and often well laminated. Shale is typically green, often weakly laminated with an anastomosing fabric, or structureless in outcrop. Some shale bands pass laterally from green to dark green to black in colour (e.g. Col. C at 4 cm). Many bands are carbonaceous. Lenticles of red silt often occur in the green shale matrix. Chert bands are very variable. Typically chert is white in colour, but some weathers red. The chert does not appear to be primary – it frequently breaks up into nodules, lenses or concretions, which distort the surrounding laminae. Some chert beds become extinct laterally, and some bifurcate, divided by silt or shale lenses. A few chert nodules cross-cut laminae, but most do not. It is possible that two generations of chert may be present.

Lateral persistence

Even the smallest shale and silt beds have remarkable lateral persistence. The best example in Figure 5.20 is the shale bed at 20.0 to 20.2 cm in Column A. This bed, which is 2 mm thick, was traced continuously across the 21 m taped, and continued for a longer distance until the end of the outcrop. Thus the thickness to length ratio is at least 1000:1 and probably very much more. All other silt and clay bands could also be laterally correlated, though in some cases with some thickness variation.

Bed fabric

Most beds show laminations. In silt units these are often well developed 1mm thick parallel laminations, marked by alternating layers of slightly coarser and slightly finer material. In shale units anastomosing patterns are much more common, and some small sigmoidal features also occur.

Many beds consist of 0.1 cm alterations between silt and clay-sized particles, often associated with a red-green colour change. In some beds (e.g. 7.7 to 11 cm, Col. D), alternating layers consist of red silty material and fine green internally-laminated clay material. In most beds which consist of a mixture of shale and silt, silt tends to occur as lenses up to 30 cm long and 2-3 mm thick, not as continuous layers. In contrast, some beds of similar thickness, such as that of 20.0 cm in Column A discussed above, are extremely continuous laterally. In thicker silty lenses, green laminations sometimes occur.

Erosional contacts are rarely seen. The bed at 11-13 cm, Column D, has a probable erosive bottom contact, with overlying green clay transgressing red laminae. At 35 cm in Column D a clear erosional contact is seen, with green clay transgressing red silt. Otherwise erosional features are very rare indeed.

Deformation

Several horizons in the rock are deformed (Fig. 5.19c), while beds immediately above and below remain intact. Deformation is generally characterized by asymmetric folding and very variable and irregular folding of no standard wavelength. Frequently folds are overturned and thrust out along or below the axial plane. Fold axes orientations do not appear to be related to any of the recognized regional deformation phases which have affected the rock (Chapter 3). The distribution of fold asymmetry and shear displacements suggests that overlying units moved west south-west. Figure 5.21 shows fold orientations. The chaotic nature of the folds and the presence of undeformed material above and below suggests that they formed during soft-sediment deformation, soon after deposition, most probably by slumping during early diagenesis. However, this is not the only possible explanation. Significantly, chert stringers and nodules are involved in the

113

Figure 5.22. Outcrop of ironstone, Ngezi River. Note lateral continuity of bands.

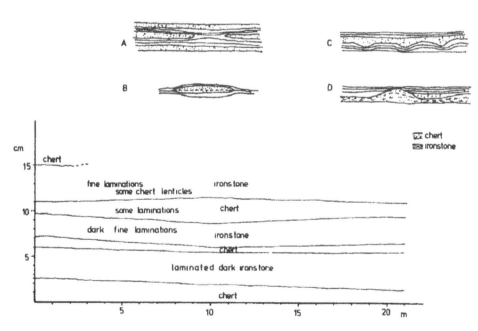

Figure 5.23. Detailed sketch of 20 m section along strike in ironstone (near tape in Fig. 5.22). A-D show examples of specific features seen in outcrop (see text).

114

folding, implying that it took place after chert formation. Chert was a very early product of diagenesis if this is a syn-sedimentation deformation. Deformed horizons are often of the order of 10-20 cm thick, over- and underlain by undeformed strata. This suggests that the chert formed during the time it took to deposit this thickness of sediment. Alternatively, it is possible that later chert exactly mimics previously deformed laminae, but many deformed chert layers appear to have been fractured during the deformation, not to have formed after deformation. Our preferred explanation is that the deformation took place some years after initial deposition when the rock was partly lithified (silicified) but mostly not indurated. The deformation may have occurred during liquefaction in earthquakes or under slumps.

To summarize, several points stand out:

1) The beds are very persistent laterally, and few erosional contacts are visible.

2) The sediment is very fine-grained.

3) The chert probably formed in very early diagenesis.

4) Soft sediment deformation probably took place.

5.6.2 *Interpretation*

The strata must have been laid down in very quiet water. Average sediment influx was probably very limited, except on rare occasions, and bottom currents very weak. The thin varve-like beds such as that at 20.0-20.2 cm in Column A were probably laid down by single unusual events. Since the rocks studied directly overlay the Black Prince Conglomerate, it is reasonable to suppose that an event which deposited a conglomerate bed in one locality may have produced a major influx of silt-laden turbid water. This would eventually settle to produce a varve-like layer away from the locus of conglomerate deposition. Such layers may have had a much higher chance of preservation in the Archaean than today, in the absence of bioturbation.

The great lateral persistence of the alternations between lithologies is probably mainly a consequence of the lack of erosion and bottom relief. The sediment deposited probably varied in grain size and thickness according to the proximity and activity of distant submarine mass flow events; at one time an active flow channel might be relatively nearby and silt would be deposited, at other times the area would be isolated and would only receive a small amount of sediment from clouds of turbid water. Within each recognized bed, reworking by very gentle bottom currents must have taken place to produce lenticles of silt and clay.

The deposit was probably laid down on a moderate slope, or in an earthquake-prone area, in which early slumping may have been a fairly common phenomenon. This is consistent with the evidence from the quartz breccias that strong topographic relief existed on land, and from the Black Prince Conglomerate to suggest significant subaqueous slopes.

5.7 THE IRONSTONE FACIES

The best exposed outcrop of ironstone in the Brooklands Formation is in the Ngezi River immediately above the shale-siltstone outcrop, which itself is above the Black Prince Conglomerate. Elsewhere in the Brooklands Formation ironstone is abundant, typically exposed as weathered-out ridges. In the Ngezi outcrop shale-siltstone appears to pass smoothly into ironstone, although the contact is partially covered by the Ngezi River. Shale-silt rock with occasional chert bands and nodules passes up-section into a much more chert-rich rock in which chert bands become laterally continuous while the shale-silt matrix passes to fine iron-rich laminae. There is no obvious break in the succession. The present-day total thickness of ironstone is greater than 50 m. Above the ironstone is a poorly exposed outcrop of mafic rock, but it was not determined whether this was intrusive or extrusive.

5.7.1 *Fabric*

The ironstone facies consists of alternating layers of 1 cm thick chert bands and 1-10 cm thick iron-rich bands (Fig. 5.22). In the iron-rich bands lateral continuity is extremely high (Fig. 5.23). Some chert bands are continuous laterally, but others break up across section into horizons of elongate lenses (Fig. 5.23A). Iron-rich microlaminae tend to wrap around the chert lenses; usually the microlaminae are not cut by the chert. However, in one example (Fig. 5.23B), a chert lens appears to have destroyed an iron-rich lamina, while the iron-rich laminae above and below have folded around the chert lens. The lens thus appears to be younger than the laminae, and to be a replacement phenomenon. Most lenses have symmetrical top and bottom surfaces, but one lens (Fig. 5.23C) has downward proturbances, possibly of diagenetic origin.

Some bands of chert may be redeposited. They appear to consist of tabular fragments of chert redeposited in the iron-rich matrix. There is no indication as to how energetic this process may have been, but the very fine-grained and undisturbed nature of the matrix suggests that the environment was very quiet, and that the chert may have been broken up by shrinkage cracking.

Erosional features are uncommon, but do occur (Fig. 5.23D). In this example the chert layer may have been exposed by erosion before deposition of the overlying iron-rich laminae in the pocket formed by the shape of the surface of the chert lens.

5.7.2 *Interpretation*

Little can be said from internal evidence about the depositional environment of the ironstones except that it was quiet, that there was very little input of clastic material and that bottom currents were very weak.

5.8 PALAEOGEOGRAPHIC SYNTHESIS

Despite the deformed and metamorphosed nature of the rocks, it can be seen from the above descriptions that there is still sufficient information preserved to attempt a rough palaeogeographic reconstruction. Each member and each facies recognized contains information: The structural setting, petrology of volcanic rocks and regional stratigraphy also have important implications.

To consider the basement first: The composition, size and shape of clasts in the coarser sediments of the Ndakosi Member strongly imply that a proximal older gneiss-greenstone terrain existed, and probably that it was being rapidly eroded and had rugged relief. Bickle et al. (Chapter 3) discuss the structural setting of the Brooklands Formation. It was probably laid down unconformably upon the 3500 Ma gneisses immediately to the East. Part of the evidence for this unconformity is sedimentological, as discussed here, but the inference is supported by regional mapping and structural and metamorphic contrast. If the 3.5 Ga gneisses form a basement, then it is reasonable to suppose that they also were the source from which the clastic sediments were derived.

116

The marked thickness variations in the Brooklands Formation, and in particular in the basal Ndakosi Member are of great interest. From the map and section (Figs 5.1, 5.2) and from Table 5.1 it may be seen that in the Ndakosi Member shallow-water (or even locally subaerial) facies predominate. This is true both in the north of the formation, where V.W. and Asci breccias occur in a setting of quartzites, pebbly sandstones, grits and banded ironstones, as well as in the south. Despite the problems of variation in tectonic strain it is obvious that a very much thicker succession has been laid down in the north, yet both successions are shallow water in facies. To compound the problem, the north appears to be in the proximal area: Clast sizes and the diversity of the sequence are greater in the north, and the few current directions available indicate transport from the north. Further, in the Brooklands Formation there are major facies changes across the locus of the Mtshingwe Fault (in contrast to the situation in the overlying Manjeri Formation, where there is less change across the trace of this fault) and most units in the Ndakosi Member are local in extent.

The simplest palaeogeographic interpretation of the Ndakosi Member is that the member was laid down as a proximal fan or fans close to the rapidly uplifted gneiss terrain to the East. On occasion, catastrophic events would have occurred, producing slumps, debris flows, rockfalls and coarse pebble conglomerates; in more normal times thick sequences of gravels, sands and silts would have been laid down. Relatively pure quartzites in the lower part of the Ndakosi may be a shallow-water or shoreline facies, but unfortunately primary structures are rare in them, and have probably been destroyed by later events. The discontinuous ironstones in the Ndakosi may represent shallow-water overbank deposits in areas protected from sand deposition. Part of the Ndakosi Member may have been deposited subaerially, part at least was subaqueous.

The enormous thickness variations in the Ndakosi Member may in part be explained by a 'fan' model, especially if part of the northern succession is subaerial, but in view of the evidence for subaqueous processes even in the north it is nevertheless likely that the Ndakosi Member was laid down on a subsiding basement of slipping fault blocks (Fig. 5.24). Such an explanation would easily explain the rapid lateral facies changes: It would imply that the regional structural regime was tensional. Possibly the Mtshingwe Fault locally reactivated an earlier structure.

This fits neatly with the next event – the eruption of lavas of the Roselyn Member, which can simply be regarded as the immediate consequence of regional extension (see later below).

The Mnene River Member also shows lateral thickness and facies changes but the contrasts are less dramatic. The most striking feature is the association of conglomerate, siltstone and ironstone. The great lateral persistence of the conglomerate unit implies that the deposit is not a shoestring (except in the very unlikely event that the present erosion surface exactly exhumes the shoestring). It is much more likely that the conglomerate forms a sheet – perhaps 10 km long and as much across. The relative lack of thickness variation along the exposure of the conglomerate suggests that total deposition was similar across the whole fan – i.e. that the sea floor where the deposition took place was smooth, with a steady slope. It is unlikely that the sea floor in the area of deposition was rugged since this would lead to local thinning or breaks in the conglomerate. However it should be noted that transport (as opposed to deposition) of the conglomerate probably took place in a distinct channel.

The characteristic lack of erosional features in the conglomerate suggests that deposition was below the level of wave action, possibly at a depth of several hundred metres. However, the large clast size implies that the flows were very competent: In other words either that there was a steep slope in the transport channel, or that the flows were thick and

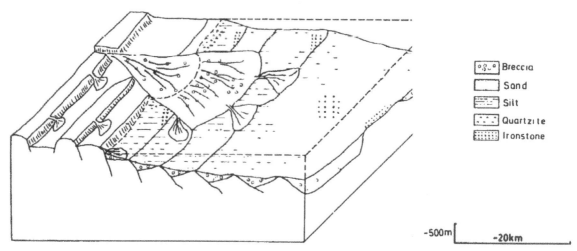

Figure 5.24. Palaegeographic reconstruction of depositional environment in Ndakosi Member, during early phase of subsidence. On land, strongly faulted terrain, with development of alluvial fans. Rockfalls and talus deposits below cliffs and steep slopes, elsewhere deposition of sand and silt. Formation of major depositional fans building out to sea, laid down on rapidly subsiding basement. Rapid transgression by sea. In quieter areas, away from fans, deposition of silts and, in very distal areas or on subaqueous high ground, ironstones.

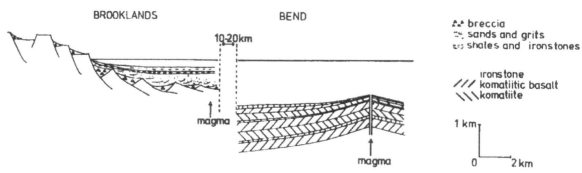

BROOKLANDS   BEND

10-20km

magma

magma

Figure 5.25. Paleogeographic relationship between Bend and Brooklands Formations. Note: This assumes that the two formations are laterally equivalent.

dense, or both. A small erosional channel cut into the conglomerate in the Lower Ngezi outcrop implies that locally and at intervals currents were indeed active, possibly a feature of deposition on the upper part of a submarine fan.

It should be noted that the conglomerate consists of very few individual flows: The whole unit could have been laid down extremely rapidly by a very small number of catastrophic events, such as the collapse of a tidal boulder bar.

The siltstones immediately overly the conglomerates. It is possible that they represent overbank levee or back-levee deposits in an upper fan to a mid-fan setting. Alternatively, the conglomerates may represent a rare event, and the siltstones may represent a more long-lived period of deposition of pro-delta muds and silts (Fig. 5.24). Slowly, as the water deepened, the deposition changed to produce banded ironstones and cherts in an offshore setting. Possibly some ironstones also formed in quiet protected shallower areas between channels, but the considerable thickness of the Mnene River Member banded ironstones implies that deposition occurred in a stable (i.e. offshore) area, not in an area where shifting delta lobes would periodically interrupt the formation of ironstones.

118

Finally, the Pemba Member consists of thick volcanic rocks and minor ironstones and cherts, probably laid down in deeper water at a time in which subsidence of the basement was more rapid than infill by volcanism or by sedimentation.

5.9 REGIONAL SETTING

Bickle et al. (Chapter 3) have discussed the structural evidence that supports the correlation of the Brooklands Formation with the Bend and Koodoovale Formations to the south-west (Orpen 1978). Both formations have had similar structural histories, and both contain komatiitic volcanic rocks interbedded with ironstones, although these are restricted to the upper part of the Brooklands Formation. If these formations are indeed chronological equivalents it is interesting to consider the implications of the correlation.

If the Bend and Brooklands Formations are chronological equivalents then it is possible that the Bend Formation was laid down in a distal setting, 10-20 km (before folding) away from the proximal Brooklands Formation. The formations may serve to delineate the margin and centre of a graben (Fig. 5.25). At the margin, rapid uplift of older terrain provided abundant detritus for debris flows and grain flows out onto the graben, with the construction of alluvial fans and deltas. Gneissic clasts are relatively uncommon, compared to other lithologies: The basement was apparently mainly of metamorphosed 'greenstones' and sedimentary and volcanic cover. At the centre, volcanism occurred, with periodic deposition of banded ironstone during periods of quiescence. The lava pile would form a submarine topographic high, so little clastic material would be deposited on it. On occasion, volcanism must have also occurred on the margin of the graben, to produce the komatiitic Roselyn and Pemba Members of the Brooklands Formation.

In an earlier section it was pointed out that, despite the great thickness variations in the Brooklands Formation, most of the sediments appear to have been laid down in shallow water. The very few palaeocurrent indicators preserved imply sediment transport from south to north. It is possible that this reflects deposition on a continuously subsiding basement of faulted blocks (Fig. 5.24), with rotational tilting. If so, the dominance of proximal shallow-water facies, the coarsening of the sequence and the increase in sedimentary diversity to the north could all be explained by a single simple model. Such rotational faulting is characteristic of an extensional setting and would be consistent with the igneous history of the sequence.

McKenzie (1978) has modelled the initiation of a sedimentary basin, and his model as applied to the Archaean has been discussed at length by Bickle & Eriksson (1982) and McKenzie et al. (1980). It is possible that the preserved sequence in the Ndakosi Member of the Brooklands Formation represents the initial subsidence of the Belingwean graben, as a result of regional crustal extension.

To conclude, the Brooklands Formation contains a wide variety of sedimentary rocks, the interpretation of which allows us to construct a model of the development of a late Archaean graben on ancient continental crust. Into this graben were deposited a variety of clastic sediments and banded ironstones, interbedded with volcanic rocks. The predictions of the McKenzie stretching model (1978) are consistent with the general sediment succession, although the model cannot be tested in detail. The model would explain the variation in the thickness of sediment laid down, and suggests the possibility that rotational faulting of the graben margin took place. If the model is realistic, it is possible that the volcanism took place as a result of stretching, the eruption of komatiites or komatiitic basalts being a consequence of regional extension. Thus the formation of the greenstone belt may be seen as a local consequence of a regional tectonic event.

To answer the questions posed at the beginning of the analysis of the Brooklands Formation:

1) The formation was apparently deposited on a subsiding, extending floor, but nevertheless a floor of continental crust.

2) The nature and thickness variations of the sedimentary rocks imply that regional extension, as suggested by the McKenzie stretching model, was the main tectonic process controlling deposition.

CHAPTER 6

The Ngezi Group: Komatiites, basalts and stromatolites on continental crust

E.G. NISBET, A. MARTIN, M.J. BICKLE & J.L. ORPEN

ABSTRACT

The Ngezi Group comprises the Manjeri, Reliance, Zeederbergs and Cheshire Formations. Although local erosional unconformity exists in the sequences, particularly at the contact between the Zeederbergs and Cheshire Formation, the sequence is essentially conformable and represents a remarkably well preserved and near continuous record of the development of the greenstone belt.

The Manjeri Formation includes roughly 100 m of sedimentary rocks, with local variation. It records the transition from the subsidence of the underlying basement to deeper water conditions. It is laid down with an unambiguous unconformity on granitic-gneiss basement in the east and on the Mtshingwe Group in the west.

The Reliance Formation includes 1 km of mafic and ultramafic pillow lavas, flows, minor tuffs and high-level intrusions. A lower komatiitic basalt member, mostly pillow lavas, flows and tuffs is overlain by a central komatiite member of ultramafic pillows and flows and an upper extrusive komatiitic basalt member.

The Zeederbergs Formation is up to 5.5 km thick and is predominantly composed of basaltic pillows and flows, associated with thin tuffs. Some komatiitic basalts occur; sedimentary rocks are almost absent apart from minor volcaniclastic deposits.

The Cheshire Formation includes roughly 2 km of sedimentary rocks and minor igneous rocks. The top is not seen. Conglomerates, ironstones, thick siltstones and limestones occur, in a very varied assemblage, most of which was laid down in very shallow water. Among the limestones are very well-preserved stromatolitic units which display a wide variety of spectacular structures.

6.1 INTRODUCTION

The Ngezi Group provides a detailed record of the development of a single greenstone stratigraphic sequence, with evidence for early subsidence (including a basal unconformity on a granite-greenstone crust), both mafic and ultramafic volcanism and the construction of a huge edifice of basaltic lavas, and then an episode of shallow-water sedimentation in which living organisms clearly played a role. The preservation of the basal unconformity on older granite-greenstone crust provides a rare example where a continental setting for an apparently typical greenstone sequence can be demonstrated. Careful

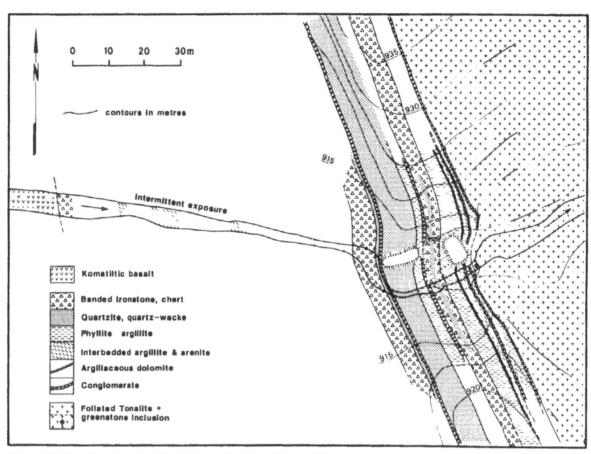

Figure 6.1. Map of type locality, Manjeri Formation (from Martin 1983).

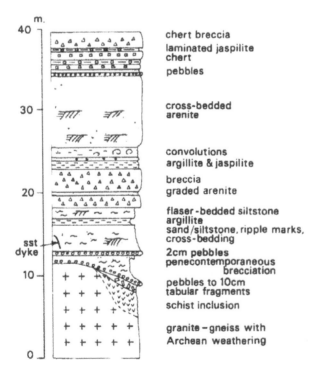

Figure 6.2. Section through the lower 40 m of the Manjeri Formation, type locality.

description of the Ngezi Group is therefore merited, not only because of the very well preserved Archaean sedimentary and volcanic rocks (see Chapter 8), but also because the succession may be used as type example of a greenstone sequence formed in a continental setting.

In this account the stratigraphy of the group is described in some detail. The group is the type succession of the Bulawayan Supergroup, and equivalent strata can be correlated across much of the Zimbabwe craton (Wilson 1979). It is thus of considerable local importance as well as being a source of information about the wider problems in Archaean geology.

6.2 THE MANJERI FORMATION

6.2.1 *Introduction*

The Manjeri Formation is the basal unit of the Ngezi Group, which forms the upper greenstones of the Belingwe Belt. The Manjeri succession is thin, but since it marks the transition from the basal unconformity below the Ngezi Group to the volcanic suite comprising most of the upper greenstones it is a succession of some interest. In very few other places is there preserved so well-exposed a contact between Archaean granite-gneiss basement and Archaean cover rocks. The formation is here described in some detail, because of its very special stratigraphic importance.

The Manjeri Formation is entirely of sedimentary origin, and ranges in thickness up to 250 m or more, though it is locally absent. It outcrops on both limbs of the Belingwe syncline. Typically, outcrop is restricted to resistant bands of quartzite and ironstone, but locally good exposure occurs, and the formation can be traced around much of the perimeter of the Ngezi Group as a series of steep low hills capped by quartzite and ironstone. In the north of the belt it has been either eliminated by younger granite intrusion, or is absent.

The metamorphic grade is typically low in the formation, except near intrusive granite. Abell et al. (1985b) have argued from stable isotopic and petrographic data that south of the type section temperatures have never been above 200-300°C at most. However, greenschist facies actinolite-epidote assemblages in adjacent Reliance Formation lavas indicate the occurrence of higher temperatures possibly localised in hydrothermal systems. The depositional fabric is well-preserved in many of the rocks and in many localities no cleavage is discernible except in the finest-grained strata. In hand specimen, sedimentary structures are often little strained. However, in the north of the belt the rocks of the Manjeri Formation are more thoroughly schistose. In the following discussion sedimentological, not metamorphic terminology is used, with the proviso that all rocks have undergone some recrystallisation.

6.2.2 *Lithology*

The formation consists for the most part of a clastic sedimentary sequence laid down initially in shallow and later in deeper water. A minor but significant component of the formation is not clastic, and includes shallow-water carbonates and ironstones.

The most important feature of the formation is the basal unconformity. The contact is well-exposed at several localities around the belt: Exposure is best at the type section of the formation, but well and differently shown at several other places. Sections through these well exposed localities are described below. This is followed by a description of regional variations and interpretation of the sedimentary environments.

Figure 6.4. Detail of basal unconformity. Right: Leucotonalitic-gneiss. Left: Basal quartz breccia passing up into clean quartz sandstone.

Figure 6.3. Outcrop of the basal unconformity to the Ngezi Group.

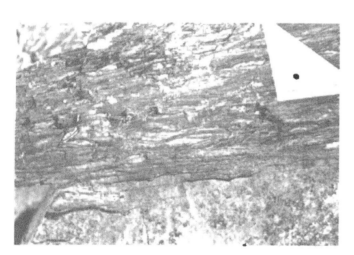

Figure 6.5. Intertidal sedimentary structures in lower part of Manjeri Formation, type section.

Figure 6.6. Underside of ripple marks, lower Manjeri Formation, type section.

124

6.2.3 *Type section: The basal contact and overlying succession*

The type section was found some years ago by Laubscher (1963), and described and illustrated by Bickle et al. (1975). It was defined by Martin (1978). Figure 6.1 is a sketch plan of the type locality, from Martin (1983) and Figure 6.2 summarises the stratigraphic column of the lower part of the type section. Notable features at base of the type section include the unconformable basement in Figure 6.3, with evidence of Archaean weathering in the underlying foliated leucotonalite, and the presence in the plane of the unconformity of a channel cut into a schist inclusion. The base of the formation is marked by a 30 cm band of subangular pebbles (Fig. 6.4) and cobbles (up to 15 cm long axis). Locally, away from the type section, clasts are much larger. Clasts are mainly of vein quartz, with some mafic igneous material and rare chromite clasts. These pebbles and cobbles appear to represent redeposited relict Archaean surface rubble. Immediately above the basal conglomerate is a thin shale band, sometimes displaying ripple marks. In the eroded channel bed this shale unit reaches 1 m thick, and consists of brecciated tabular shale fragments, ranging up to 20 cm long and several cm thick. It is clearly a penecontemporaneous breccia and resembles a 'mudflake conglomerate.' The upper part of the shale band, away from the channel, contains well-developed ripple marks and lenticular bedding.

Above the shale is a 50 cm thick sandy bed (2 mm grain size), passing into a poorly sorted pebble bed containing subangular quartz pebbles set in a coarse arenaceous matrix. This pebble bed contains a number of thin (0.5 cm) clay flasers, and passes up into 3 m of varied silts and minor sands, in beds 2-20 cm thick. Arenaceous and clay lenses are common in this unit, and a variety of sedimentary structures is displayed (Figs 6.5-6.11), including climbing ripple laminations, cross-bedding features (with trough cross-bedding and small-scale cut-and-fill structures), flaser-bedding and a 30 cm long neptunian dyke (Fig. 6.9).

These varied silts and sands pass into an 80 cm thick massive band of sandy dolomite, which is structureless apart from thin argillaceous stringers at the base. Above the massive band are finely laminated red silts and argillites, locally displaying flaser bedding, ripple marks and cross-bedding. These silts are iron-rich and in thin section contain quartz, opaques (haematite) and altered clay minerals.

On the hills to the north and south of the type section, the lowest 3 m of the Manjeri Formation consist of quartzitic and arkosic arenite lying on the basal conglomerate and pebble beds. Thin pebble beds are common in the arenite.

The first banded ironstone unit in the succession occurs above the fine ferruginous silts. The ironstone consists of alternating thin (0.5-1 mm) cherty bands, containing abundant fine silt-sized quartz crystals in a siliceous matrix, and iron-rich bands containing abundant opaque oxide grains. The rock is very finely laminated and locally contains brecciated tabular clasts of the same material. This intraformational conglomerate is similar to that observed at the base of the succession.

Above the ironstone is a 40 cm thick arenaceous unit, with 5 cm thick beds. These beds are graded and show parallel laminations and occasional load structures. Further laminated argillite and brecciated banded ironstones follow. The ironstone consists of alternating 0.5 mm thick laminae of iron-rich and iron-poor silt-sized grains, cemented in a siliceous matrix. Fine-grained carbonate minerals are an important constituent of the matrix in iron-rich bands. Pyrite crystals are common in some bands. In the ironstone tabular clasts of apparently penecontemporaneous breccia may locally be seen, and in thin section cracked and broken laminae are apparent, overlain by undeformed laminae. This brecciation appears not to be a tectonic feature. Rather, it seems to be a consequence

Figure 6.7. Sedimentary roll, Manjeri Formation, type section.

Figure 6.8. Sandstone (some dolomitic) and shale interbeds, Manjeri Formation, type section.

Figure 6.9. Neptunian dykes (at point of pen), Manjeri Formation, type section.

of syneresis cracking or, more probably, subaerial dehydration. The close alternation between laminated ferruginous argillite and banded ironstone suggests that similar processes produced the two rocks.

The alternating ferruginous argillite/ironstone passes into a 10 cm thick pebble band, with 5-10 cm long clasts of vein quartz, chert and ironstone fragments. Some clasts range up to 30×10 cm, and the pebble band can be traced across the outcrop for at least 30 m. Above the pebble band siltstone returns, with fine cross-laminations and flaser bedding. This siltstone consists of mainly quartz clasts, and minor mica, set in a matrix dominated by carbonate minerals with subsidiary silica. It passes up to an 8 m thick arenite with some cross-lamination.

At the top of the arenite is a further pebble bed, containing well-rounded quartz pebbles 3 cm in diameter, and tabular clasts of chert and ironstone (4×1 cm in 2 dimensions), set in an arenaceous matrix. Overlying the pebble band is an 80 cm thick coarse arenaceous quartzite bed, passing into chert and jaspilite. The next 5 m consist of alternating red jaspilite and dark waxy-lustre chert. Jaspilite is typically very fine-grained and locally contains intraformational breccia: Tabular clasts up to 30 cm long and several cm thick, possibly a result of storm rip-up (Fig. 6.10).

The top of the section in Figure 6.13 consists of laminated argillaceous material, in 1 cm bands. Some bands are very iron-rich.

Above this there is a break in the section of 20 m. Above the break, the formation continues with a thick greywacke sequence of alternating argillite and coarse arkose. It is possible that the gap in the section also contains arkosic rocks, which weather more rapidly than quartzite and ironstone.

The main succession of greywackes consists of a series of 20-40 cm thick arenaceous beds interbedded with argillite. Arenaceous bands are sometimes graded from 1 cm diameter pebbles to fine sand, and some show parallel lamination and poorly developed cross lamination. Some arenite beds have slightly irregular bottoms and appear to have eroded underlying argillite during deposition. Occasionally stringers of argillite occur in the arenite and may be rip-up features or injection structures from underlying mud. Probable flame structures exist.

The more structureless arenite beds are continuous across outcrop, but some thinner sandstone bands are more irregular. Some of the latter have small-scale lenticular features, ripple marks and flasers.

In the lower part of the greywacke succession around 70% of the succession is arenaceous and 30% thin argillaceous layers. Toward the top of the succession this proportion is reversed and the rocks become predominantly fine-grained.

The highest unit of the sedimentary succession is a seven metre thick band of sulphide-facies ironstone. This band displays complex folding of component quartz and iron-rich layers, and it is possible that part at least of this folding may have been penecontemporaneous and represent synsedimentary slumping.

Above the ironstone are 4 m of altered mafic material and then komatiitic flows and pillow lavas of the Reliance Formation.

The entire sequence is 105 m thick. It is described further by Martin (1978,1983).

6.2.4 *Eastern margin of the syncline*

The Manjeri Formation outcrops both to the north and south of the type section on the eastern margin of the syncline.

North of the type section the Manjeri Formation extends to Zwamperi Hill, 8 km north-east of Zvishavane, where it is intruded by younger granite. For much of this

Figure 6.10. Penecontemporaneous brecciation in banded chert and jaspilite, Manjeri formation. Top: Continuous lamination. Bottom: Clasts, 5-20 cm long, randomly orientated.

Figure 6.11. Small erosion structure, possibly showing early induration of lower cross-bedded unit, Manjeri Formation, type section.

distance the Manjeri forms the capping to a low range of hills on the boundary of the greenstone belt, and exposure is mainly restricted to resistant quartzite and ironstone. A persistent horizon of sulphide-facies ironstone occurs at the top of the formation, usually exposed as a siliceous or earthy gossan.

Near Zvishavane, in the cutting of the Bannockburn-Zvishavane railway line 4 km southwest of Zvishavane is a well-exposed section (Fig. 6.12) through the Manjeri Formation, including the basal unconformity. The basement in this locality consists of weathered foliated leucotonalite, outcropping in the bed of a small stream below the railway. The outcrop is 3 km from basement granite dated by Moorbath et al. (1977) at 3495 ± 120 Ma (Rb-Sr whole-rock isochron; see Table 2.2). The contact between basement and Manjeri Formation is exposed over a distance 20-30 m (Fig. 6.14). The contact is irregular with a surface relief of about 1 m. Some of the 'pockets' in the contact contain breccias of tabular clasts of siliceous mudstone, thought to be of very local derivation, possibly after subaerial mudcracking. Quartz breccia is also very common at the contact, in lenses up to 20 cm thick set in 'pockets' in the surface of the unconformity.

The typical contact rock away from 'pockets' in the basement is a 10-25 cm thick bed of arkosic sand containing quartz pebbles up to 1 cm in diameter.

Above the contact is ironstone, with 0.1 cm thick bands of chert and iron-rich rock. Locally the ironstone breccias contain ironstone clasts up to 10 cm long in a quartzite matrix. Field relationships strongly imply that this breccia is penecontemporaneous, as it is overlain and underlain by undisturbed layers and the deformation was probably not tectonic. Some ironstone horizons are rich in pyrite and locally ferruginous conglomerate occurs with ferruginous, quartz and orthoclase clasts set in a siliceous and chloritic matrix. This in turn passes into alternating beds of dark ferruginous arenite and banded jaspilite, locally brecciated (probably a syn-sedimentary feature).

Above the alternating arenite/jaspilite comes a thick succession of arkose sandstone and quartzite, with occasional chert stringers and quartzite bands. The arkose typically contains alkali feldspar and quartz clasts, often subrounded to well rounded, set in a siliceous or chloritic matrix. The quartz clasts are sometimes strained but often not, and appear to have come from a granitic or gneissic terrain. Chert beds occur occasionally in the arkose, and part of the section is very iron-rich.

The arkose passes up into beds of sandstone and breccia, the latter containing angular fragments of quartzite and jaspilite fragments up to 15×5 cm in outcrop, set in a matrix of arkosic sand. This passes into a coarse conglomerate containing tabular clasts of jaspilite (20×4 cm in outcrop) and more rounded clasts of chloritic schist (5×4 cm), set in a matrix of arkose sand. Beds of ferruginous arkose also occur.

Above this is a break in the main section, but a quarry 25 m off the line of section contains arkose sandstone and coarse conglomerate. This passes into a dominantly conglomerate rock, containing clasts of finely banded arkose sandstone (clasts up to 20×4 cm), quartzite and dark chert, chloritic fragments and common granite clasts. These granite clasts are often well rounded and up to 15×10 cm. They are relatively fresh and unmetamorphosed, and are not foliated.

After a further break in the succession, the arkose sandstone becomes dominant and is only locally conglomerate. Some sandstones have significant calcite in the matrix.

Most arkose sandstones are nearly structureless in outcrop, with 50 cm thick massive beds. Occasionally these beds have irregular bases, possibly representing contemporary erosion of the underlying surface during deposition of the bed above.

Grading is not visible, although this may be partly a consequence of the unweathered state of much of the outcrop.

The top of the Manjeri Formation is marked by a 'sulphide-facies' ironstone horizon.

129

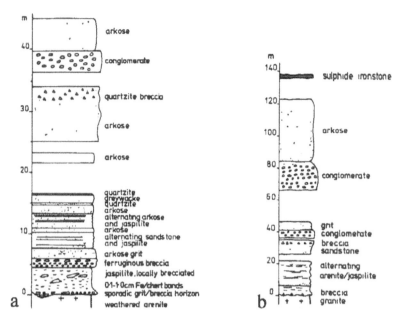

Figure 6.12. Manjeri Formation near Zvishavane, in railway cutting. a) Detail of lower 40 m. b) Overall succession.

Figure 6.13. Deeper water facies, upper Manjeri Formation, type section.

This horizon contains grey-white, 5 cm thick chert bands and red jaspilite bands, and weathered earthy ferruginous crud. Much of the outcrop is highly deformed, in contrast to the low state of deformation of the rest of the section. Part of the deformation may represent syn-depositional slumping.

Above the sulphide-facies ironstone there is no further exposure, but the nature of the topography suggests that the subsoil rocks are komatiites of the Reliance Formation, as is confirmed by regional mapping.

6.2.5 *South of the type locality*

South of the type section the resistant banded ironstone and quartzite of the Manjeri Formation continue as a broken range of hills to Mt. Manjeri, after which the formation is named. South of Mt. Manjeri, between the peak and the Bannockburn-Vukwe railway line, the formation consists of lenses of quartzite, ferruginous phyllite and limestone.

South of the Bannockburn-Vukwe railway line the Manjeri Formation forms the crest of the Rupemba and Pemba ranges. From the railway line to the Mtshingwe Fault the Manjeri Formation is considerably disrupted by vertical bedding plane shears as well as the later north-west striking sinistral faults. Here the formation is underlain by the sedimentary and volcanic succession of the Brooklands Formation. The contact is not exposed, but mapping and the structural evidence implies that the Brooklands Formation was deformed and eroded prior to the deposition of the basal Manjeri sediments.

Along Rupemba Range and as far as the Mtshingwe Fault the Manjeri Formation stratigraphy comprises quartzites overlain by carbonates, some stromatolitic, and capped by the prominent magnetite-jaspilite banded ironstones. Interbedded with these lithologies, but less well exposed, are chlorite-rich pelites, chlorite rich siltstones and more rarely conglomerates. The latter have been seen in only two localities underlying the main ironstone band and contain rounded 5-20 cm clasts of chert, banded ironstone, chloritic pelite and vein quartz in a chlorite quartz grit matrix. There are also a few small bodies of intrusive serpentinite. Both the carbonates and the ironstones exhibit tight and isoclinal folds and locally brecciation. This deformation makes it impossible to determine how much of the repetition of beds within this stratigraphy is tectonic. At the south-west end of the range where quartzite and carbonate are complexly interbedded this repetition in part may be sedimentary. Locally it is difficult to discriminate Manjeri quartzites from quartzites in the underlying Brooklands Formation.

The basal quartzites are derived from relatively pure orthoquartzites. The carbonates range from brecciated chlorite-dolomite rocks to pure limestones. The less deformed carbonates include rocks with 1 cm thick chert and calcite bands as well as finely laminated calcite-dolomite rocks of possible algal origin. In one locality, 0.5 km due east of Rupemba peak, the carbonate band is little deformed and contains well preserved algal stromatolites described in detail by Martin et al. (1980) and Abell et al. (1985b). In this vicinity the carbonate-ironstone sequence is apparently repeated by shearing but the upper limestone band contains isoclinal folds and sedimentary structures are not seen. The banded ironstones outcrop along the ridge crest. Mostly these contain alternating magnetite, jaspilite and chert bands. Locally these pass laterally into weathered brecciated zones, probably derived from sulphide-rich ironstones. The upper of the two 5 m bands exposed in the Ngezi is stained by sulphides. Between Rupemba and the railway line the banded ironstone is overlain by a thin, poorly exposed conglomerate containing 5-10 cm banded ironstone clasts in an iron-rich pelitic matrix.

South of the Mtshingwe Fault the Manjeri Formation is less deformed than along the Rupemba Range. In the North the sequence again consists of quartzite, overlain by grey

Figure 6.14. Contact between Manjeri Formation and basement near Zvishavane. Left: Manjeri Formation. Right: Basement.

Figure 6.15. Cross-bedding in arenite, Manjeri Formation, western reference section.

limestones capped by the magnetite-jaspilite-chert-banded ironstone. The quartzite contains scattered rare clasts of jaspilite, dark chert and vein quartz in a 2-4 mm pure quartz sand matrix. Clasts are mostly 1-2 cm in size but with rare banded ironstone boulders as large as 50×10 cm. It is at the base of this quartzite that the contact with the Brooklands Formation is most nearly exposed. Here thin banded ironstones of the Mnene River Member strike at right angles to the overlying Manjeri quartzite and, although chloritic pelites are sheared on the contact, the undeformed state of the quartzite and the banded ironstones suggest that this is not a major shear zone. 600 m south of the Mtshingwe Fault the quartzite pinches out and is not seen further south. The overlying carbonate becomes more deformed and only scattered dolomite-chlorite outcrops are mapped to the south. The jaspilite-magnetite banded ironstones retain their thickness and locally overly a conglomerate of 5 cm rounded quartzite clasts in a chlorite quartz-grit matrix. Above the banded ironstone, a conspicuous conglomerate is developed. The first outcrops are seen about 700 m south of the Mtshingwe Fault and the band maintains a thickness of 100 to 150 m for 4.5 km to the south. The conglomerate contains subangular to subrounded chert and banded ironstone clasts, mostly between 1 and 5 cm but up to 10 to 20 cm in certain areas. The matrix of iron-rich chlorite pelites and quartz grits forms between 30 and 50% of the rock. The top of the conglomerate is not well exposed but locally it is capped by thin dark cherts. Both the conglomerate and the underlying banded ironstone are well developed on Pemba hill.

South of Pemba both the banded ironstone and the conglomerate thin significantly. Where the Manjeri Formation crosses the Mnene-Lou Estate track the conglomerate is absent and the banded ironstone reduced to thin cherts and iron rich pelites. South of this in an area of poor exposure the thin cherts, banded ironstones and chlorite-rich pelites can be followed for a further 5 km to where the greenstone belt is cut by Chibi Batholith. South of Pemba the angular discordance between Manjeri and Brooklands Formation is small and although the Reliance Formation with abundant spinifex komatiites is distinctive it is not possible to distinguish between the cherts, banded ironstones and pelites of Manjeri and those interbedded with the mafic volcanics of the underlying Pemba Member of Brooklands.

6.2.6 *Western margin of the syncline*

On the western margin of the syncline the Manjeri Formation is of similar thickness and lithological content, although it was probably deposited several tens of kilometres distant from the eastern deposits. The principal difference is the lack of thick limestone and quartzite beds on the western side although thin quartzite beds are common and a number of thin limestone horizons outcrop in the south-west area.

In the Bend area, Manjeri Formation lithologies, especially chert, ironstone and phyllite, are often difficult to distinguish from Bend or Koodoovale Formation lithologies. However the limestone, quartzite and sulphide-facies ironstone characteristic of the Manjeri Formation are absent or very rare in the Mtshingwe Group. Furthermore, detailed mapping (Orpen 1978) reveals a well developed angular unconformity at a locality six km north of Bend Mine (Fig. 6.16). At this locality the basal sedimentary unit exhibits rapid lateral facies variation from dolomite, chert/ironstone, limestone, phyllite, conglomerates with subrounded 2 cm chert and ironstone pebbles to quartzite comprised of moderate to well rounded quartz and chert grains. The limestone exposed due south of the Python Kop is in direct contact with ironstone striking with Bend Formation rocks. At Cockroft Ridge and Bobbejaan Kop (Fig. 6.16) several outcrops expose the contact between the basal Manjeri quartzite and underlying folded and brecciated ironstone. In both localities the quartzite becomes coarser and more conglomeratic with mainly subrounded, ironstone clasts adjacent to the basal contact. The Bend Formation ironstone horizon traced from Belingwe Peak forms a small palaeo-high on the unconformity surface and the quartzite horizon can be traced from Bobbejaan Kop to a level 20 m topographically below and abutting the projecting top of the Bend ironstone unit on its north-western side. The variable basal Manjeri horizon is overlain by oxide or sulphide facies ironstone, and generally much less well exposed sandstone and phyllite. Relatively well exposed ripple-marked and current-bedded arenaceous rocks are found on the east flanks of Bobbejaan Kop.

Similar lithologies are found to the north of Belgian Hill and to the south in the axis of the Dube anticline. The Koodoovale Formation is overlain by moderately continuous

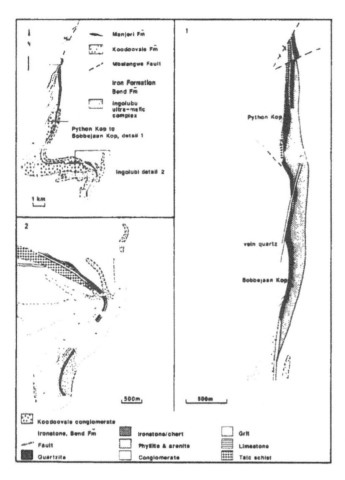

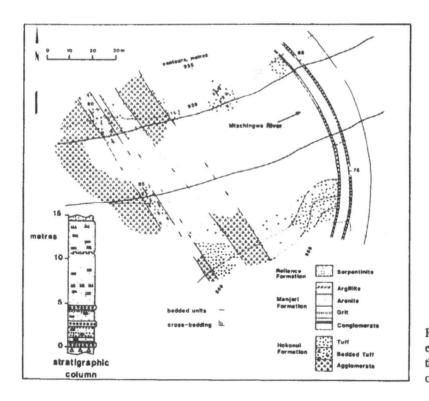

Figure 6.16. Outcrop of the Manjeri Formation in the Bend area. Top left is general map showing trace of unconformity. Right is unconformity outcrop in Python Kop-Cockroft Ridge-Bobbejaan Kop-Belingwe Peak region. Note rapid lateral facies change. Bottom left is Ingolubi outcrop showing Manjeri Formation intruded by sills of the Ingolubi complex.

Figure 6.17. Map of river section, reference locality, Manjeri Formation in the Mtshingwe river valley on the west of the belt.

sulphide-facies ironstone but a 100 m dolerite sill intruded much of the contact and poor exposure precludes demonstration of an unconformable relation.

On the southern boundary of the greenstone belt, near Mnene mission, the Manjeri Formation has been destroyed by the Chibi Batholith, which has intruded into the Zeederbergs Formation.

Near Mberengwa the Manjeri Formation is poorly exposed and all that is visible is a band of resistant quartzite, ironstone and conglomerate a few metres wide. Near the Mtshingwe Fault the formation is only a few metres thick and consists of argillite and ironstone.

6.2.7 *Mtshingwe River outcrops – Reference type section*

Two good outcrops of Manjeri Formation occur north and south of the Mtshingwe River, separated by the Mtshingwe Fault.

South of the Mtshingwe River is an abandoned slate quarry cut into fissile shales of the Mtshingwe Formation. The formation here rests on foliated Hokonui Formation rocks, but it is difficult to determine the exact contact. A partial section is exposed in a valley approximately 50 m west of the slate quarry.

The underlying strata are felsitic agglomerates of the Hokonui Formation. Associated with the agglomerates are possible conglomeratic units, but exposure is poor.

In the outcrops on the southern bank of the Mtshingwe River a pebble conglomerate band is taken as the base of the Manjeri Formation. This conglomerate contains clasts of the underlying felsitic volcanics, set in a matrix of quartz and chlorite. Clasts average 2×2 cm in outcrop, but some range up to 10 cm or more in length.

Above the conglomerate are fine silty and sandy beds, containing clasts probably derived from the Hokonui Formation, which pass into red siltstones. This siltstone correlates with the fine exposure in the slate quarry of ripple marked silts. In the quarry, surfaces up to 10 cm long are displayed showing very regular ripple marking. Ripples average 2 mm deep, with a half-wave length of 1.5-2 cm. The siltstone is 3-5 mm thick.

Above the ripple-marked siltstone is sandstone and then quartzite. The quartzite contains clasts of strained quartz and micropegmatite, as well as common feldspar clasts. Locally the rock is arkosic and may have been derived from both Hokonui and granitic terrains.

North of the Mtshingwe Fault the contact between the Manjeri Formation and the Hokonui Formation is exposed in the Mtshingwe River valley, and a section through the basal 15 m of the Manjeri Formation is exposed. This section has been designated a reference type section by Martin (1978). Figure 6.17 is a map of the outcrop.

At the base of the section the contact with the underlying Hokonui Formation is exposed. Measurement of dip surfaces has revealed an angular unconformity of 10-20° on this surface. The unconformity is shown in Figure 6.17, from Martin (1983).

The Hokonui Formation, locally highly weathered, passes into a basal conglomerate consisting of subrounded to subangular clasts of felsite and jaspilite up to 8 cm long, passing into 1.5 m of sandstone and breccia, sometimes showing fining-upward cycles. Above this is a second conglomerate horizon, again containing angular clasts of felsite and jaspilite and interbedded with sandy bands. This passes into parallel-laminated quartz-sand and then a further conglomeratic horizon, with alternating bands of conglomerate and coarse sand. Above this is parallel laminated sandstone.

A conglomerate horizon overlies the sandstone, and some red calcareous clasts are visible in this conglomerate, which passes into parallel laminated sands and grits, often calcareous. Two horizons of red nodules (now weathered out) are visible in the calcareous

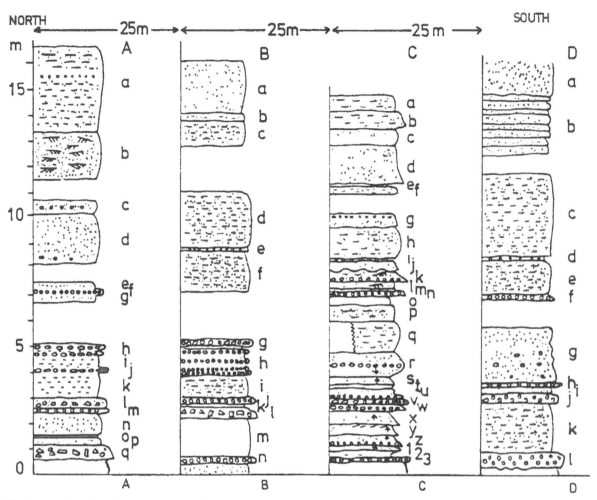

Figure 6.18. Stratigraphic sequences in the Manjeri Formation, western reference section.

Column A: a – parallel laminated quartzite; b – cross-bedded quartzite; c – nodular horizon; d – coarse sand-pebbly sand (clasts up to 15 cm); e – grit; f – conglomerate; g – coarse sand; h – conglomerate; i – parallel laminated grit; j – conglomerate clasts 2-5 cm of Hokonui Formation; k – parallel laminated grit; l – conglomerate clasts 2-5 cm; m – breccia, clasts 2-5 cm; n – grit; o-q – jaspilite clasts up to 8 × 1 cm.

Column B: a – quartzite; b – quartzite; c – parallel laminated quartzite; d – parallel laminated grit; e – red nodular clasts; f – parallel laminated grit; g – conglomerate; h – alternating conglomerate/coarse grit; i – parallel laminated sand; j – conglomerate; k – sand; l – breccia including clasts of Hokonui Formation and some jaspilite; n – conglomerate. Base not seen.

Column C: a – quartz sand; b – parallel quartz grit; c – parallel laminated quartzite; d – quartzite; e – argillite; f – parallel laminated sand; g – calcareous nodules, sand with red silt lenses; h – parallel laminated sand; i – nodular horizon; j – sand (erosional base); k – sand; l – sand, cross-bedded; m – grit; n – graded pebbly grit; o – parallel laminated fine sand; p – parallel laminated fine sand; q – parallel laminated fine sand (off section); r – grit, coarsening up at base; s – sand, fining up; t – quartzite, parallel laminated at top; u – sand; v – conglomerate/breccia; w – conglomerate; x, y – parallel laminated sand/grit, fining up; z – breccia, fines up; 1 – sand; 2 – silt; 3 – 1 cm conglomerate.

Column D: a – reddish quartz grit; b – bedded orthoquartzite; c – reddish parallel laminated grit; d – nodular horizon; e – parallel laminated coarse sand; f – conglomerate: red clasts; g – variable quartzite/coarse sand, locally with pebbles 1 cm; h – conglomerate; i – coarse sand; j – breccia with Hokonui Formation clasts; k – parallel laminated grit; l – conglomerate with subrounded Hokonui Formation fragments, 5 × 3 × 3 cm. Base weathered Hokonui Formation.

136

sandstone. Above this the rocks become predominantly quartzose, typically with parallel lamination, although in the north of the section a 2 m thick bed has very well developed cross-bedding (Fig. 6.15).

The uppermost part of the exposed section is quartzitic, but has ferruginous horizons. Figure 6.18 shows four measured sections 25 m apart through the sequence in the area of the reference stratotype.

North of the Mtshingwe River the Manjeri Formation becomes less well exposed and consists of fine grained arenites, poorly exposed argillite (typically weakly cleaved), local limestones and resistant bands of chert and banded ironstone in the upper part of the formation. The cherts are normally massive white or pale gray rocks, and ironstones have distinct beds 1-10 cm wide. Conglomerate is well exposed five kilometres along strike from the reference type locality, north of the Mtshingwe Fault. Here a small stream cutting through the Manjeri Formation has formed a cliff face composed of conglomerate. The pebbles appear mostly to be derived from the underlying Hokonui Formation and are up to 15 cm across. Three to six kilometres north of the Mtshingwe River the chert-ironstone horizons are replaced by vein-like quartz bodies. North of this the rocks are characteristically deformed, and the outcrop of the Manjeri Formation is restricted to lensoid bodies of ironstone, chert and quartzite, and vein quartz. The Manjeri Formation appears to be absent in the north-west of the mapped area, where the Reliance Formation appears directly to overly the basement. Poor exposure characterizes this area but the exposure of thin chert horizons in the underlying Hokonui Formation suggests that chert or quartzite on the unconformity would be observed, if present.

6.3 DEPOSITIONAL FACIES IN THE MANJERI FORMATION

Several distinct facies of deposition may be identified in the Manjeri Formation, and their analysis provides some evidence as to the nature of the initial subsidence phase of the Ngezi Group.

6.3.1 *Basal facies*

The basal facies (Fig. 6.21) is typically conglomeratic to sandy, though more rarely it includes chert or ironstone. A basal horizon dominated by subangular clasts of vein quartz is characteristic in the eastern outcrops of the Manjeri Formation, above the ancient granite-gneiss basement. This basal conglomerate probably represents material derived from pre-Manjeri surface rubble lying on the old granite-gneiss surface. Under present-day weathering the old granite-gneiss basement is typically covered by vein-quartz rubble and coarse sand, subject to wind-winnowing and sheet-flood erosion. Comparable material constitutes the basal Manjeri beds. The outcrop pattern suggests that the initial relief was fairly subdued as the base of the Manjeri shows few major indentations into the granite, although small local channels occur which were probably cut as subsidence (and uplift elsewhere) began. Quartz sand layers in the basal Manjeri probably represent early sandy beaches; there is no evidence for a cliffed shoreline. The 3.5 Ga granite-gneiss terrain was probably erosionally mature, of low relief and deeply weathered (as is shown by the preserved fossil weathering profile in some outcrops). Small channels must have existed and may have been preferentially eroded into schist inclusions in the basement, as is seen on the type locality. Locally however (e.g. near Zvishavane, where the conglomerate contains varied clasts up to 20 cm long), a source of coarse clastic material existed.

137

Figure 6.19. Conglomerate, Manjeri Formation.

138

Figure 6.20. Oblique aerial view showing the low topography of the Reliance Formation, east limb of the greenstone belt syncline. View looking north from the vicinity of Rupemba-Manjeri range right to distant centre. Zeederbergs Formation left. Shabani ultramafic in distance on right (east).

Above the Brooklands, Bend and Hokonui Formations, the Manjeri Formation typically contains locally derived clasts and matrix. In the Bend area thin chert-clast conglomerate and chert-rich quartzite overly a rocky coastline produced by resistant ironstone horizons in the underlying Bend Formation (Fig. 6.16). Above the Hokonui Formation the Manjeri Formation contains a conglomerate of felsic volcanic clasts. In the Brooklands area basal chert-clast conglomerates are mapped in the rather disrupted geology of the Rupemba range. To the south around the Pemba range a thick and coarse chert-clast conglomerate (Fig. 6.19) overlies a basal ironstone suggesting local progradation of a shore line into rapidly deepened water. Above the Koodoovale Formation the base of the Manjeri Formation is a persistent sulphide-facies ironstone overlain directly by komatiitic basalts of the Reliance Formation. However south of the Koodoovale Formation outcrop in the Dube anticline conglomerate, quartzite and carbonate facies reappear above Bend Formation rocks.

6.3.2 *Shallow water*

Above the basal facies a clastic and shallow water facies (Fig. 6.22) in most sections, except where the Manjeri Formation is absent, overlies the Koodoovale Formation or in the central and southern parts of the Brooklands area. Notable features of this facies include:

1) The presence of conglomeratic and pebble beds. These beds were probably laid down as wide thin lenses (see Fig. 6.18). Only infrequently do they have markedly erosional bases, and they are associated with a variety of other deposits, ranging from sandstone to jaspilite.

Sometimes associated with pebble beds, and common in all sections are intraformational conglomerates and breccias. It is most probable that these conglomerates and breccias are the product of cracking and slumping of clay and iron-rich layers during dessication. During later flooding the mud chips would be swept into nearby sediment traps to form mud-breccia horizons. The presence of neptunian dykes in the sandstones supports the concept of rapid deposition during flooding. During a flood, sheets of pebbles and sand would be laid down, and mud chips eroded and deposited.

2) Ripple marks, cross-bedding and fining upwards sequences are common in the variable sands and argillites. The common flaser and lenticular bedding indicates a very shallow water fluvial, lacustrine or intertidal environment, and the variability of the deposits suggests that widely fluctuating currents were reworking the available material.

The absence of common erosional features in the restricted outcrops suggests that if the environment were fluvial, then deposition must have been on a rapidly subsiding basement.

The intraformational conglomerates suggest that there was periodic dehydration of the muds, most probably a consequence of subaerial exposure. This could suggest a seasonal lacustrine environment, or that the muds were overbank deposits in a fluvial system. However, the sedimentary structures are best interpreted as being dominantly intertidal deposits, probably laid down in small fans and deltas. Some deposits may be subaerial, some subaqueous, but most were probably laid down at the coastline by small streams with irregular flow and common flash floods. In the south rather larger rivers may have existed. Locally, but not generally, in the north of the belt (on the granite-gneiss) scarps may have delimited the coast, producing angular clasts.

3) Calcareous and dolomitic deposits are common in the Manjeri, associated with the variable clastic facies and in the south and east as distinct limestone reefs, with stromatolites (described by Martin et al. 1980). Major limestone bodies appear only to occur

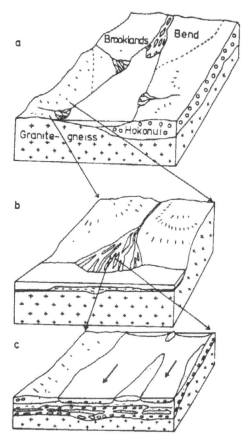

Figure 6.21. Possible reconstruction of the depositional environment of the basal Manjeri Formation development of a transgressive sequence with beaches and small fans growing on drowned coastline.

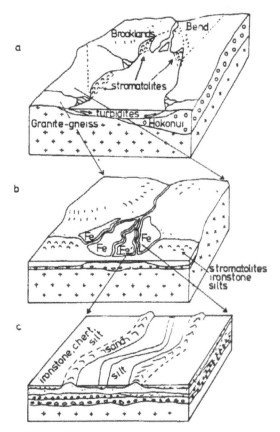

Figure 6.22. Possible reconstruction of the depositional environment of the shallow water facies of the Manjeri Formation, with variable shallow-water clastics, ironstones and stromatolites. Very shallow-water but detritus free conditions of deposition suggest that the ironstones may have been laid down in marshy or overbank areas, subject to storm rip-up.

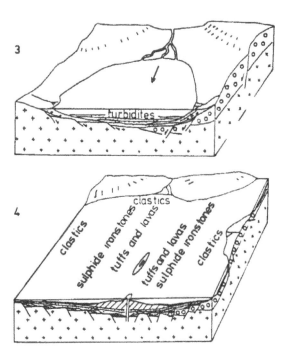

Figure 6.23. Possible reconstruction of the depositional environment of the deeper water clastic sequence in the Manjeri Formation, with turbidites and eventually, onset of volcanism.

above an older greenstone basement. The limestone may have formed as bioherms in bays in areas of low relief, above the less resistant volcanic and sedimentary rocks of the lower greenstone. Where limestone is absent, carbonate-rich sands are common and calcite or dolomite often forms the matrix in arenaceous rocks. It is clear that carbonate minerals formed an important component of Archaean sedimentation, at least in this region.

4) Locally within the shallow water deposits are stringers and bands of banded ironstone, of oxide and more minor carbonate facies. In places intraformational breccias are common in the ironstones and the rocks are often closely associated with ripple-marked and flaser-bedded argillites, silts and sands. It seems likely that some of the ironstones were laid down in very shallow water and were probably subject to periodic dessication. They may have been the normal deposits of sediment-starved shallow-water, occasionally interrupted by incursions of comparatively rapid deposition of clastic material (e.g. in floods or storms). The normal fine lamination may represent seasonally or tidally-controlled slight changes in environment. It is probable that the ironstones represent the normal deposits in sediment-starved shallow to medium depth environments in Archaean time. Intraformational breccias in the ironstone suggest that in part the rocks may have been overbank deposits, periodically subject to erosion and redeposition by migrating channels (possibly tidal). Possibly some of the ironstones may have formed on tidal flats.

6.3.3 Deeper water

In the east of the belt, near and south of Zvishavane, the majority of the Manjeri Formation consists of a 70 m thick greywacke sequence. This sequence is not seen elsewhere, probably as a result of poor exposure. In the type section (Bickle et al. 1975; Martin 1978), the greywackes are well-bedded arkoses with argillaceous intervals. The sandy beds, 1 to 20 cm thick, are typically continuous over outcrop and in places show structures such as grading, cross-lamination and parallel lamination as well as basal erosional features and rip-up structures. Occasionally load structures may be seen.

At the top of the Manjeri Formation is a sulphide-facies ironstone, of great lateral persistence. From the evidence of the underlying strata this was probably deposited in deeper water (Fig. 6.23) or after submergence of the detrital source. It is overlain by volcanic rocks, and was most probably formed in a deep, quiet setting close to a growing lava pile, which may have significantly altered the chemistry of the water.

To summarize the initial subsidence of the upper greenstones: It appears to have been shortlived from the nature of the sedimentary rocks deposited (100 m of sediment) but with fairly rapid deepening with few erosional events. The deepening facies sequence implies that the rate of infilling by sedimentation was much less than the rate of subsidence. At the beginning of subsidence the surrounding terrain was probably erosionally fairly mature: This may in part account for the inability of sediment supply to keep with subsidence. In essence, then, the bottom dropped out.

6.3.4 Current directions and the geometry of the depositional environment

Limited palaeocurrent data are available from the Manjeri Formation. A detailed description of the measurements and corrections for unfolding is given after the discussion of the Cheshire Formation which follows (Fig. 6.44). It should be noted that unfolding introduces some uncertainty. The available information from the Manjeri Formation indicates transport parallel to the main synclinal (D_g2) fold axis of the greenstone belt. Note that

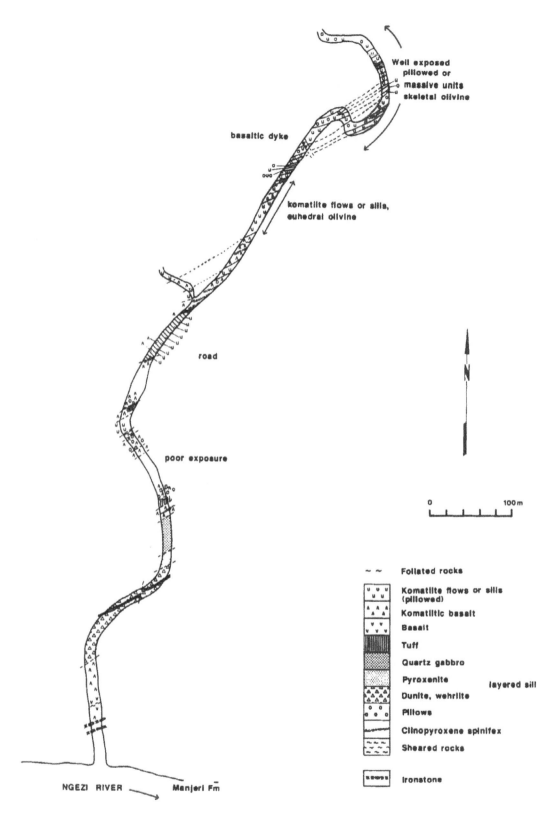

Figure 6.24. Map of type section of Reliance Formation, after Martin (1983) and this work. Solid ornament is rodingite-like dykes.

current directions in Figure 6.44 are plotted relative to an unfolded greenstone belt with the D_g2 axis trending north-east. Thus the current direction at the Pemba locality trends parallel to present strike and the D_g2 fold axis.

This information does not in itself constrain the geometry of the depositional environment for which a proper facies analysis is required. The limitations imposed by the two dimensional sections of the Manjeri Formation preclude a unique interpretation of facies variation. Salient points are: 1) The Manjeri Formation thins to minor chert, phyllite or ironstone or is completely absent in the north of the belt and in the far south-east; 2) major quartzite horizons occur in the central part of the eastern margin and thin discontinuous quartzite occurs along the western margin; 3) major limestone occurs in the south-central part of the eastern outcrop (Rupemba to Pemba) and thin limestones occur in the south-west exposures above the Bend Formation; 4) lithological content of the basal Manjeri Formation strongly reflects local provenance. Such variations could be explained by inundation during a progressive craton-wide transgression, moving north to south across the belt or by subsidence along a north-northwest trending trough (or graben) centred along the main synclinal axis of the Ngezi Group Greenstone Belt. The coincidence of the current directions with the fold axis is of interest. Martin (1983) argues that the greater degree of metamorphism in the northern part of the belt implies greater uplift in this area. He interprets the transition from shallow water clastics in the centre of the area to thin cherts or complete absence of sediment in the north of the area as representing a transition from the margins to the centre of a localised trough-like depository. The rather rapid local variations of facies in the south-west and south-east suggest that in the south the story is more complex. A definite model is not possible at present. The success of the correlations by Wilson et al. (1978) suggests that compilation of Manjeri facies variation over a much wider area of the craton may be possible and this might allow reconstruction of depositional environments not possible from the Belingwe data alone.

6.3.5 Depositional models

The reconstructions presented in Figures 6.21-6.23 are thoroughly speculative. They rest to some extent on the very weak evidence, discussed above, that the formation was laid down in what was a local focus of deposition in a broader transgression. If the formation instead records a simple transgressive change across the area a similar model could be constructed with only one exposed source area.

Initially the basal conglomerates and pebble beds together with variable sands and silts may have been deposited by braided rivers and streams which built out small fans and deltas over the pre-Manjeri surface. Many small consequent streams may have discharged into the Manjeri sea or lake.

As the water deepened and became more extensive, stream deltas probably built up to some extent and a beach-intertidal facies may have developed. The common intraformational conglomerates and breccias were probably eroded from lagoonal and overbank deposits where argillite and ironstone were forming in the absence of vegetation. Dessication would be common in this environment, causing mud-cracking.

Algal mats and stromatolites may have formed in lagoons between stream deltas.

In deeper water, turbidites deposited the well-bedded greywackes as the Manjeri basin widened. Finally, much of the perimeter of the basin must have been submerged before volcanism began, and sulphide-ironstone formed on the floor of the starved basin. Only locally does Reliance Formation appear to rest directly on basement. These places, scattered around the perimeter of the belt on both margins, may represent outstanding land promontories at this stage.

143

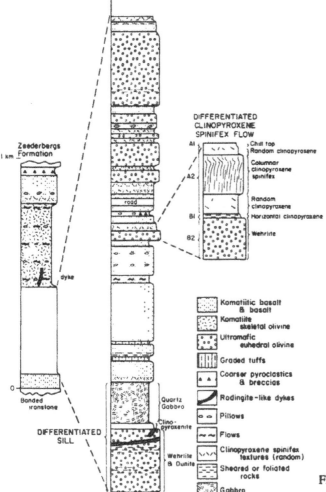

DIFFERENTIATED
CLINOPYROXENE
SPINIFEX FLOW

A1 — Chill top
Random clinopyroxene

A2 — Columnar clinopyroxene spinifex

Random clinopyroxene

B1 — Horizontal clinopyroxene

B2 — Wehrlite

Zeederbergs Formation
1 km

dyke

0

Banded ironstone

DIFFERENTIATED SILL

Quartz Gabbro

Clino-pyroxenite

Wehrlite & Dunite

road

Legend:
- Komatiitic basalt & basalt
- Komatiite skeletal olivine
- Ultramafic euhedral olivine
- Graded tuffs
- Coarser pyroclastics & breccias
- Rodingite-like dykes
- Pillows
- Flows
- Clinopyroxene spinifex textures (random)
- Sheared or foliated rocks
- Gabbro

Figure 6.25. Stratigraphic column of type section, Reliance Formation (from Nisbet et al. 1977).

Figure 6.26. Columnar jointing, lower mafic member, Reliance Formation, Ngezi River.

6.4 THE RELIANCE FORMATION

The Reliance Formation conformably overlies the Manjeri Formation. It consists of a diverse suite of igneous rocks, most of which are extrusive and many of which are highly magnesian. These latter include both komatiites and komatiitic basalts. Most of the rocks are metamorphosed and altered, with the metamorphic minerals reaching low green-schist assemblages, but in some places remarkably fresh rocks occur, and olivine and clinopyroxene are abundant and rarely calcic plagioclase feldspar and relict glass occur. Figure 6.20 shows a general view of the topographic expression of the rocks.

The formation is stratigraphically coherent although significant shear zones do occur in it. The many and ubiquitous facing directions obtained all point inwards to the synclinal axis of the belt. We thus conclude that it is not allochthonous: The Reliance Formation rests conformably on the Manjeri Formation, which rests on basement. The upper contact with the Zeederbergs Formation also appears to be conformable.

The formation was described in some detail by Nisbet et al. (1977). In this account we attempt to supplement rather than to duplicate that description, although for convenience the geology of the type section is summarised again.

6.4.1 Stratigraphic succession

The type section of the formation is located in the bed of a small intermittent stream informally named Umvobo spruit west-south west of Vukwe beacon. The formation can be divided into a lower member 400 m thick of basalts and komatiitic basalts, a central unit of 400 m of komatiite and an upper mafic unit, with approximately 200 m of mafic lavas and some tuffs. Figure 6.24 is a map of the type section and Figure 6.25 a strati-graphic column up the section, modified from Nisbet et al. (1977).

6.4.2 Lower mafic member

The contact between the Manjeri Formation and the Reliance Formation is best seen in drill core (courtesy of Union Carbide Ltd.) from close to the western type section of the Manjeri Formation, which shows thin beds of quartz grit intermingled with tuffs of the basal Reliance Formation, implying that some clastic sediment was deposited after the onset of volcanic activity. Otherwise, only limited volcaniclastic deposits occur in the Reliance and Zeederbergs Formations. In general, however, the volcanic succession of the Ngezi Group is very poor in clastic sediment, implying either that most land was rapidly covered in lava or that eruption took place so rapidly that the input of clastic material was in contrast of very minor proportion.

The lower part of the member includes massive tholeiitic basalt, overlain by 60 m of komatiitic basalt containing both hollow and solid megacrysts of magnesian pigeonite up to 0.3 mm across set in a felted tremolitic groundmass, with minor feldspar microlites. Overlying the komatiitic basalt in the type section is a 90 m thick differentiated igneous body of overall composition in the range 15-20% MgO. This body includes serpentinite after wehrlite, overlain with sharp contact by pyroxenite and then gabbro and finally quartz gabbro. It is possible that this body is a thick magnesian flow. It can be traced for many kilometres along strike away from the type section. Above this are 80 m of poorly exposed mafic rocks, including a thin graded tuff band and komatiitic and tholeiitic basalt, overlain by a massive komatiitic basalt flow, and then komatiitic basalt pillow lavas. Above this in the type section is a complete differentiated lava flow with an olivine-rich base and a thick clinopyroxene spinifex zone. Further komatiitic basalts

Figure 6.27. Lava tubes infilled with quartz, lower mafic member, Reliance Formation.

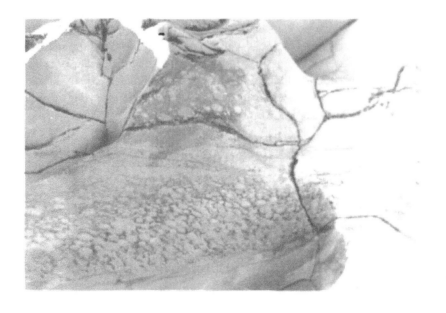

Figure 6.28. Graded accretionary lapilli tuff, Reliance Formation, Ngezi River. Knife for scale in lower centre.

overly this, displaying a variety of clinopyroxene spinifex textures (Fig. 6.29), and the top of the unit is defined as the contact with the first pillowed komatiites, above this.

Away from the type section the lower mafic member can be traced around much of the perimeter of the belt. Locally, columnar jointing occurs, for example in the Ngezi River on the east side of the belt (Fig. 6.26). In the southwest, above the Bend and Koodoovale Formations, pillowed spinifex textured and massive basalts and komatiitic basalts occur, up to 500-600 m thick. Figure 6.27 illustrates the pillow lavas. Lava tubes and cavities infilled with quartz are common. About 150 m above the base of the Reliance a 1-2 m thick tuff is present and can be traced for several kilometres, with the best exposure at GR 044187. Much of this layer is an accretionary lapilli tuff of komatiitic basalt composition consisting of small fine grained ellipsoidal and spherical lapilli, often graded, 0.5-1 cm in diameter (Fig. 6. 28). In these rocks, lapilli and fiamme textures suggest subaerial accretion, in phreatoplinian events.

6.4.3 *Central komatiite member*

The central komatiite member is dominantly ultramafic, including an impressive extrusive komatiite suite. The komatiite pillows and flows are generally highly weathered but readily identifiable, especially after flash floods. The pillows are small (Fig. 6.30), usually 20-50 cm across with narrow selvedges and young consistently towards the centre of the syncline. Many flows, typically about 3 m thick, have been recognised. In the field flows tend to have a rubbly base, a more massive centre and a rubbly top passing to pillows. Locally, chill zones can be seen at the top and bottom of flows. Lava toes can sometimes be identified. The rocks' mineralogy is mostly totally altered (except sometimes for relict clinopyroxene and very rarely, olivine), with pseudomorphs after small euhedral but embayed olivine hoppers, quench olivine skeletons, clinopyroxene microlites and groundmass after glass (Fig. 6.31). Spinifex olivine outcrop is very rare in the Reliance Formation. In the SASKMAR locality (Nisbet et al. 1987) near Zvishavane, aligned and random spinifex plates occur, in rocks which contain fresh relict igneous olivine and clinopyroxene. Here too, quench chills on flow margins are well developed.

In the south-west of the belt the ultramafic rocks of the Reliance Formation are well exposed in the Klipspringer area in the Bend region (Orpen 1978), where a number of ultramafic dykes and sills cut a series of ultramafic lava flows. In this area are developed superb columnar structures, olivine spinifex textures and amygdales. The columns are very well-formed hexagonal prisms up to 20 m long with short axes 50-80 cm across. Columns contain roughly 80% euhedral pseudomorphs after olivine, 5% stubby euhedral clinopyroxene laths and 15% once-glassy groundmass.

In some rocks crackled flow margins can be recognised. These rocks resemble ultramafic agglomerates formed of dark green irregular globules 2-15 mm across, composed of a fine mass of altered once glassy material. Amygdales occur averaging 2-6 mm in diameter. Some show evidence for the presence of a gaseous and a liquid phase, the latter now represented by zeolites.

6.4.4 *The upper mafic member*

The upper part of the Reliance Formation consists of a variety of komatiitic basalts, basalts and tuffs, which are for the most part rather poorly exposed. The upper contact of the Reliance Formation with the Zeederbergs Formation is placed arbitrarily at the base of a range of hills of basalts (often with over 50% SiO_2): This topographic feature is distinctive and can be traced around much of the belt. A distinctive horizon of tuffs and coarser grained volcaniclastic rocks occurs at this stratigraphic level around much of the belt. There is a certain amount of overlap between the rock types composing the Reliance and Zeederbergs Formations, and komatiitic basalts do occur in the Zeederbergs Formation. The change from Reliance to Zeederbergs volcanism is one of volcanic product and any hiatus in activity is unlikely. The tuffs are well developed near the N.A. mine on the west of the belt (Fig. 6.32).

6.5 THE ZEEDERBERGS FORMATION

The Zeederbergs Formation rests with apparent conformity on the Reliance Formation,

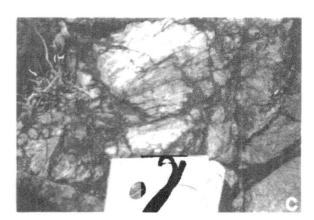

Figure 6.29. Clinopyroxene textures in Reliance Formation komatiitic basalt.
a) Clinopyroxene spinifex needles, lower mafic member, type section. b) Alternating layers of randomly orientated and columnar clinopyroxene spinifex, lower mafic member, type section. c) Sheaf of radiating clinopyroxene needles, lower mafic member, type section. d) Thin section textures. Note hollow clinopyroxenes, seen both end on and along length. Laths 1-2 mm long, or more.

and comprises a pile of mafic rocks, mostly extrusive, up to 5.5 km thick. In places it is very well exposed as basaltic pillow pavements in the Ngezi River (Fig. 6.33). In the south and centre of the belt the deformation and metamorphism are of low grade; further north and in the aureole of the Chibi Batholith amphibolite occurs. The type section of the formation is along the Ngezi River on the east of the belt.

6.5.1 *Field relationships and outcrop*

As stated above, the boundary between the Reliance and Zeederbergs Formation is to some extent arbitrary, and set at a distinct topographic break which can be traced around much of the belt. Martin (1978) has described the basal contact with the Reliance Formation, in the railway cutting near Dadaya Station, which shows pillowed komatiitic basalt of the Reliance Formation separated from tholeiites of the Zeederbergs by a thin tuff band. In the south of the area the contact is poorly exposed but appears to be conformable. In one area in the south-west, however, the Mberengwa Fault has put Zeederbergs rocks in faulted contact against lavas of the Bend Formation. In the south, a major dolerite sill close to the base of the Formation can be traced for 12 km from the Mberengwa Fault eastwards until it is lost in the amphibolites of the Chibi aureole. It is underlain by a fine grained basaltic tuff up to 4 m thick which is considered to make the base of the Zeederbergs Formation.

The bulk of the Zeederbergs Formation is the thick pile of mafic pillow lavas and flows (dominantly tholeiitic but also komatiitic), minor tuff bands, minor agglomerates, minor volcanoclastic sediments and rare felsites. Some serpentinites occur in the formation but it is not clear whether they are of intrusive or extrusive origin. Minor clastic and chemical sediments occur in the formation near the top.

The upper contact with the Cheshire Formation is for the most part a sharp transition into a succession dominantly composed of sedimentary rocks, with minor volcanic rocks. In some places the transition is into ironstone, elsewhere into conglomerate and in a few areas volcanic rocks grade into an intermingled suite of lavas and sediments.

6.5.2 *Texture and petrography*

The most characteristic Zeederbergs rock is basaltic pillow lava. Pillows range in size from 0.2 to 2 m or more long (Fig. 6.34). They give ubiquitous and unanimous facing directions into the core of the syncline. Pillow margins are generally thin and well-defined: In some places distinctive lava tubes and toes can be identified. Figure 6.33 illustrates a typical pillow outcrop. Very extensive pavements of pillow lavas occur in the Ngezi River section on the east side of the belt.

Pillows show a variety of features. Vesicles are rare, but associated with pillows are common lava tubes showing drainage cavities infilled with quartz, and quartz-filled voids are common in pillows (Fig. 6.36) especially in the upper parts. A characteristic feature of the pillows is the large number of pale weathering spheroidal structures (Fig. 6.37). Many of these spheroidal structures have a more 'felsic' rim zone. Coalescence of these spheroids is common. Their internal mineralogy is an interlocking felt of tremolite, chlorite, plagioclase and sometimes quartz. The origin of these spheroids is not understood; they may be an immiscibility phenomenon, but perhaps more likely they are rounded fragments of hot pillow margin similar to those described from modern mid-ocean ridges by Walker et al. (1979). Pillow breccia is ubiquitous, both between individual pillows and as layers. In thin section the lavas typically show a variety of intergranular-intersertal and hyalocrystalline textures. In some rocks a fine subophitic

149

Figure 6.30. Textures in ultramafic rocks. a-b) Pillow lavas, central komatiite member, Reliance Formation. c) Flow unit, with rubbly top (to right), massive interior (centre) and rubbly base (poorly exposed, left). Black lines indicate centre region of flow. Central komatiite member, Reliance Formation. d) Columnar jointing in ultramafic rock, Klipspringer peridotite, south-west of belt, Reliance Formation.

intergrowth of pyroxene and altered plagioclase occurs. Others show subradiate plumose bundles of clinopyroxene intergrown with altered plagioclase. Pillow margins preserve holohyaline textures in now altered rock. Some rocks display clinopyroxene spinifex texture.

Probably intrusive dolerite sills occur in the Zeederbergs Formation, but are difficult to identify in the field. A minor part, probably less than 5% of the formation, is composed of tuffaceous rocks, sometimes highly sheared: Indeed, in the absence of primary textures it is often difficult to decide whether a foliated band is a shear zone which has preferentially followed a tuff band, or a shear zone which incidentally happens to look like a tuff band and is misidentified as such. Most of the Zeederbergs Formation is little strained and uncleaved, and it is likely that strain during folding has been preferentially taken up by shear zones. Many tuffaceous bands in the well exposed Ngezi River section exhibit graded or small-scale cross-bedded units. In thin section, tuffs are typically very fine grained, with small ragged clinopyroxene phenocrysts and a few plagioclase phenocrysts. Agglomerate occurs occasionally, with clasts of hyalocrystalline rock and small once-vitric shards. Very few terrigenous clastic deposits occur in the main part of the formation.

6.5.3 *Metamorphism and alteration*

The rocks have undergone very variable degrees of alteration and metamorphism. In the bulk of the formation the dominant assemblage is one of small clinopyroxene granules set in a groundmass of fine clinopyroxene or fibrous tremolite-actinolite, albite, quartz, sphene and, significantly, minor epidote. Other rocks have undergone less alteration and occasionally calcic plagioclase microlites are found in association with little altered clinopyroxene and a very fine groundmass which is nearly isotropic. In contrast, in some areas extensive calcite deposition has occurred and the rocks are heavily carbonated. In general, in the bulk of the outcrop area of the formation the degree of recrystallisation is exceedingly variable, some rocks being thoroughly altered while within a few hundred metres little altered rocks occur. This variability may reflect the varying chemical and thermal environments of hydrothermal recharge and discharge systems.

In the north of the formation, there is a gradual increase in metamorphic grade, with schist common in the 'Snake's head' belt. The amphibolite is pale-green, epidote is ubiquitous. In the south, along the margin of the Chibi Batholith, hornblende occurs.

6.6 THE CHESHIRE FORMATION

The Cheshire Formation is the uppermost stratigraphic unit of the Ngezi Group. It is dominantly sedimentary, with only minor intercalations of mafic and felsic volcanic rocks. The bulk of the formation is made of siltstones associated with conglomerates and minor sandstones, tuff beds, breccias, ironstones and a very important although volumetrically minor unit of limestone. The bulk of the clastic succession is of sediment derived from a mafic volcanic terrain, and the formation shows great lateral facies variations (Fig. 6.38).

The contact with the underlying Zeederbergs Formation is not well exposed but over most of its length can be defined to within a few metres or less. The strike of the contact is parallel to bedding in the Cheshire Formation (when bedding can be seen); it is also parallel to bedding in the Zeederbergs Formation as well as that can be determined; and is parallel to the Zeederbergs-Reliance contact. In this sense the Cheshire Formation is

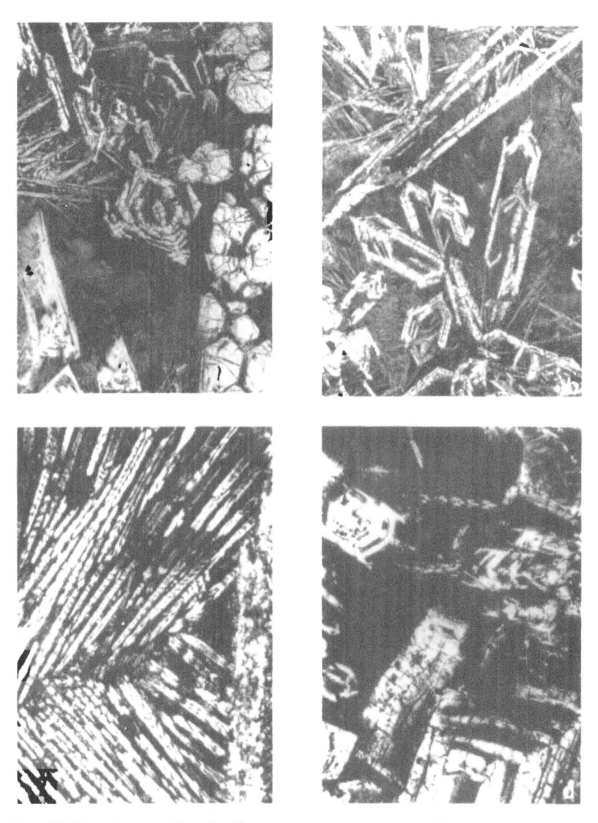

Figure 6.31. Thin section textures in komatiites. Hopper olivine crystals range up to about 0.5 mm across or more. d) Shows skeletal laths of clinopyroxene and a single altered olivine lath.

considered to be broadly conformable with the Zeederbergs Formation, although it contains much material apparently eroded from a source area laterally equivalent to the Zeederbergs terrain (see below).

6.6.1 *Lithological variation*

There is a great lithological variation in the Cheshire Formation. Along much of the eastern margin of the formation the Zeederbergs Formation passes up into conglomerates. The conglomerates are up to 500 m thick and are in places very massive. They are well developed around Zeederbergs Siding (Figs 6.39, 6.40).

Further south, the conglomerates give way to a basal succession of sulphide-facies banded ironstone and phyllites (once siltstones), which underly lenses of conglomerate that locally thin laterally to lithic sandstone. Along the southern and south-western flanks of the formation, sulphide-facies ironstone is the basal unit of the formation, and passes up into a succession of siltstones, sands (sometimes with ball and pillow textures) and tuffs, with intercalated concordant altered mafic rocks and dolerite. On the western side of the formation a limestone band appears and the Zeederbergs Formation is overlain by sandstones, then thick limestone, which passes up into siltstones. In the centre of the belt, south of the Ngezi River, the succession is yet more complex. In a minor syncline west of the main axial core of the belt, Zeederbergs lavas pass up into siltstones and then bands of limestones; in the main syncline the basal sediments are a thick lens of conglomerate which passes up to siltstone, then more conglomerate. North of the river, the basal facies are thin silts and thick limestones in the subsidiary western basin, and thin silt passing to thick limestone to silts in the main basin. Further north, the limestone disappears in gossanous banded ironstone which passes to siltstone.

This very great variation in basal facies presumably reflects a complex and not necessarily synchronous transition from the Zeederbergs Formation to the Cheshire Formation, with a complex local topography at the interface.

The core of the Cheshire Formation is also somewhat variable. In the south, conglomerate and siltstone pass up to a prominent horizon of arkose sandstone and grit and then to a succession of alternating banded ironstones and siltstones and some quartzite. Further north, siltstones predominate, and north of the Ngezi River quartzite bands are found. The top of the Cheshire Formation is not seen. The total remaining thickness of the formation is up to 2.5 km.

6.6.2 *Conglomerates*

The conglomerate is a clast-supported rock, with well rounded clasts ranging from 1.5-20 cm in diameter (Fig. 6.40). Imbrication, size grading and other sedimentary structures are rare and commonly absent. Most clasts are basaltic and were most probably derived from a terrain closely similar or laterally equivalent to the preserved part of the Zeederbergs Formation. A small proportion of the clasts includes fragments of tuffs, cherts and vein quartz, as well as quartzite.

Variolitic textures are developed in some clasts; others appear to be derived from pillow lavas or flow tops. In thin section clasts often show the feathery pyroxene and altered plagioclase texture typical of basaltic pillows. Some clasts display small scale structures in amphibolites possibly after clinopyroxene spinifex.

153

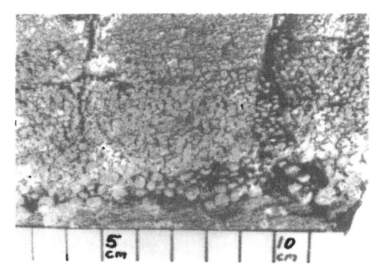

Figure 6.32. Tuffs, upper mafic member, Reliance Formation, N.A. mine.

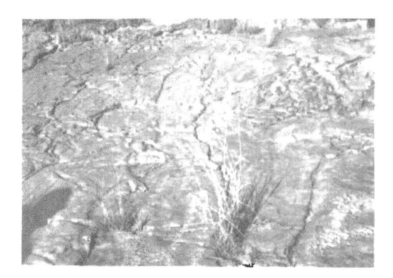

Figure 6.33. Basalt to magnesian basalt pillow pavement, Ngezi River, in Zeederbergs Formation.

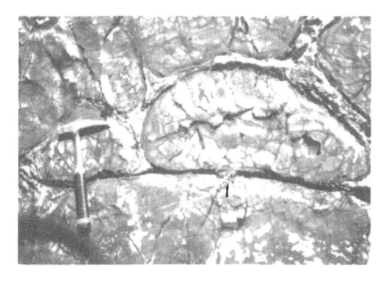

Figure 6.34. Zeederbergs Formation pillow lava.

6.6.3 *Siltstones and sandstones, quartzites and ironstones*

Much of the Cheshire Formation is composed of assorted yellow and pink weathering siltstones and argillites, and lenses of coarser sands. The siltstones and argillites typically display a parallel lamination which is often of sedimentary origin, although in some areas cleavage is commonly well-developed in the rocks and bedding may in places be rotated into the cleavage plane. Rarely, in well-exposed outcrops, sedimentary structures are apparent, and ripple marks can be seen. Most of the rocks are probably of very shallow-water origin.

In places, siltstones are interbedded with sandstones and conglomerates. In the south of the area, pebbly mudstone lenses occur, the pebbles being elongate clasts of siltstone. Sandstones are very variable, including commonly volcaniclastic material. Some arkose occurs. In places, transitions can be seen from sands to conglomerates which locally develop sand lenses.

Many of the better preserved siltstones and sandstones show evidence in thin section for the former presence of volcaniclastic material, including shards, now altered.

Quartzites are typically devoid of primary sedimentary features. Quartz clasts are up to 2 mm across and set in a matrix of cryptocrystalline quartz with minor chlorite and sericite. Ironstones include gossanous, sulphide facies rocks along the margins of the belt. In the extreme south-west of the outcrop, the Agincourt Gold Mine was set up on this band. In the centre of the belt, cherts and iron-rich bands occur, interbedded with argillites.

6.6.4 *Limestone member – Stratigraphic setting*

The Cheshire Formation contains very extensive and thick limestone beds, displaying a wide variety of textures. The bulk of the limestone is laminated or shows brecciation; in some places lamination is developed as stromatolitic structures.

In the south of the region of limestone outcrop, 2.5 km south of Mumburumu is a thick (up to about 400 m) limestone body, for the most part either massive or with parallel laminations and cherty bands. Near the top of the hill small stromatolitic structures are visible, some of which have conical cross-section. Worst (1956) analysed some samples of the limestone and reported a high quality calcite deposit, with 2-4% Fe_2O_3, Al_2O_3 etc., 2-4% SiO_2 and less than 1% MgO.

Further north, in the western subsidiary syncline to the main Cheshire structure around Mumburumu, limestone outcrops as bands roughly 10-20 m thick. Near Mumburumu small columnar branching stromatolites occur. Further north, large domal structures occur, but are in places heavily silicified. Associated with the limestones are cross-bedded silts and sands, some calcareous, often showing ripple marks, festoon cross-bedding, penecontemporaneous slump structures and liquefaction features. Pseudonodules are common. Conglomerates also occur, including rocks with angular clasts up to 3×1 cm of apparently locally derived laminated and rippled siltstones. Some conglomerates may be mudflake deposits.

The best outcrops of stromatolitic limestones occur north of the Mtshingwe Fault in the Ngezi River. Here most of the western subsidiary syncline is limestone, much of which displays stromatolitic structure, parallel laminations, or is a breccia. The limestones and associated interbedded rock of the subsidiary syncline are best exposed in Macgregor Kloof, described by Martin et al. (1980) and by Abell et al. (1985b). The geology of the limestone unit has been described in some detail by these authors and is thus only briefly summarised here.

At the Macgregor Kloof locality the subsidiary syncline is approximately 400 m

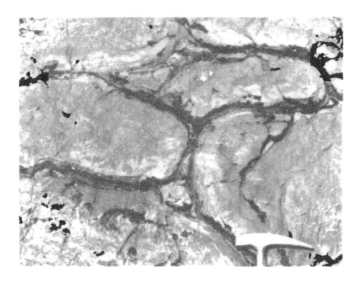

Figure 6.35. Pillow lava, Ngezi River. Note altered margins and presence of leucocratic ocelli. Zeederbergs Formation.

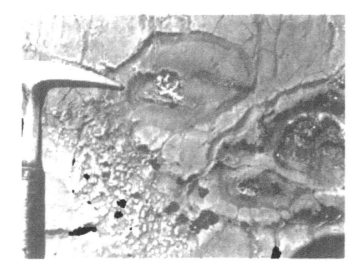

Figure 6.36. Quartz-filled void in pillow, Ngezi River, Zeederbergs Formation.

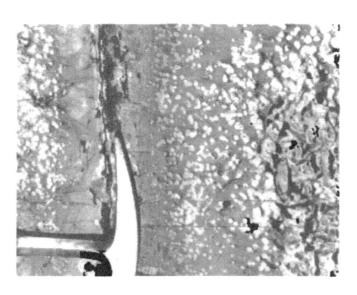

Figure 6.37. Spheroidal structure in pillow lava, Ngezi River, Zeederbergs Formation. Hammerhead is on pillow margin.

across and the limestone member is composed of a number of limestone bands interbedded with fine-grained clastic sedimentary rocks. Above the lavas of the Zeederbergs
Formation are siltstones and sandstones up to 100 m thick; on the eastern side of the
subsidiary syncline these pass into brecciated and laminated limestone, while on the west
of the syncline the siltstones pass into two to three bands 5-30 m thick of stromatolitic
limestone interdigitated with siltstone, back to siltstone and then to the main stromatolitic
band described by Martin et al. (1980). The centre of the syncline is again siltstones, circa
10-20 m thick.

The siltstones display very shallow-water structures, including symmetrical ripple-
marks, with wave-lengths about 5-6 cm and present amplitudes of 0.25 cm. Some display
trilete syneresis or mud-cracks. Other nearby siltstones show polygonal mud-cracks and
possible halite casts. Associated with the siltstones are sandstones and pebble-bands
displaying a variety of sedimentary structures including herring-bone cross-bedding and
parallel lamination. Fining upward sequences can be recognised. In many places chan-
neling can be seen, with pebbles collecting as lag in irregular erosional scours. Sandy-
pebbly units are 0.2-2 m thick. Some conglomerate also occurs, in the lower part of the
unit on the west with subrounded clasts up to 20×10 cm of grit, chert, stromatolitic
limestone and fine parallel laminated siltstones. Small chert pebbles occur in the matrix.

The limestones themselves include massive breccias and stromatolitic units. The
breccia forms a thick single bed on the eastern limb of the syncline in the Macgregor
Kloof locality. It does not persist throughout, and stromatolites occur in places. The
breccia contains clasts from 2-20 cm across, typically angular to subrounded. Clasts are
made of limestone and chert. Some are stromatolitic. The matrix is medium-grained,
recrystallised limestone.

6.6.5 *Stromatolitic limestones*

The lithology of the stromatolitic beds (Fig. 6.41) has been described in detail and
profusely illustrated by Martin et al. (1980). To summarise briefly, 33 individual beds of
stromatolites can be recognised in the Macgregor Kloof outcrop. Most of these when they
can be traced persist along the whole of the exposed strike of several hundred metres.

Martin et al. recognised 22 cyclic events in this sequence; the lower part of each event
being represented by laminated limestone or dolomitic limestone with minor chert and
argillaceous material. These lower zones comprise the bulk of the stromatolitic succession. The thin middle zones often display evaporitic textures, including radiating crystal
structures. The upper zones, also thin, show good lamination in limestone, often producing nodular or domical stromatolites. Brown-weathering dolomitic limestone and detritus are common in this zone.

Martin et al. (1980) interpreted these cyclic events as representing evaporative and
flooding periods in the history of a shallow coastal lagoon or embayment in which the
stromatolites were constructed.

Morphologically, the stromatolites are very diverse. Figures 6.41 and 6.42 show some
examples: The structures are described in detail by Martin et al. (1980) and Abell et al.
(1985b).

6.6.6 *Metamorphism and alteration*

For the most part, the Cheshire Formation has been metamorphosed in low greenschist
facies. Cleavage is sporadically developed in argillites. No detailed study has been made
of the metamorphic mineralogy of the pelitic assemblages. Abell et al. (1985a, b) have

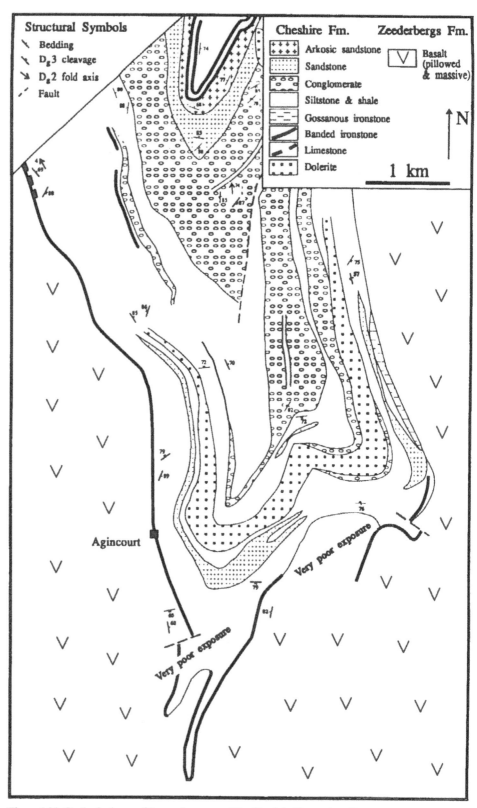

Structural Symbols

- Bedding
- D_g3 cleavage
- D_g2 fold axis
- Fault

Cheshire Fm.

- Arkosic sandstone
- Sandstone
- Conglomerate
- Siltstone & shale
- Gossanous ironstone
- Banded ironstone
- Limestone
- Dolerite

Zeederbergs Fm.

- Basalt (pillowed & massive)

N

1 km

Agincourt

Very poor exposure

Very poor exposure

Figure 6.38. Geological map of the southern part of the Cheshire Formation. Note that exposure around southern closure of Cheshire and Zeederburgs Formations is mostly very poor.

used stable isotopic data in an attempt to constrain the thermal history of the limestones, and concluded that they had had an extremely gentle history. The results were consistent with volatile loss at 200°C or less. This is supported by the brown colour of the kerogen in some samples, and by evidence from step-heating experiments, which show that ancient nitrogen is given off above 250°C (Gilmour, pers. comm.). The extremely well-preserved textures in the carbonates also suggest little metamorphic recrystallisation. Some rocks appear still to display primary and early diagenetic micrite, unrecrystallised to the limit of optical resolution.

6.7 PALAEOENVIRONMENT

The lithological diversity of the Zeederbergs-Cheshire contact invites comment about the palaeoenvironment. The salient points are that in one area there was ironstone of sulphide facies being deposited (a facies also associated elsewhere in the belt with the onset or cessation of active volcanism), in other areas massive lenses of conglomerate were being constructed, most probably in large fans up to 5-10 km long and 500 m thick, while in yet other regions apparently quiet deposition of very shallow water and intertidal siltstones occurred. Amongst all this were local protected regions where clastic sedimentation was so suppressed that stromatolites could grow with only rare influxes of detritus.

All this was most probably not synchronous in the sense that the contact most probably does not represent a time horizon, especially as tuff bands and lavas occur in the lower Cheshire succession. Yet in a wider sense, Walther's facies law would imply that these various facies were indeed being laid down synchronously.

Figure 6.43 illustrates what the early Reliance setting must have looked like. The basal Cheshire setting is shown in Figure 6.44. In some areas active volcanism probably continued after the initiation of sedimentation (and is represented by tuffs and fine volcaniclastics in the lower Cheshire succession). On high subaqueous levels starved of sediment, sulphide facies ironstone probably formed, for example on the slopes of recently extinct volcanoes close to hydrothermal systems, or possibly precipitated directly from a more reduced Archaean ocean. Elsewhere, the rugged volcanic topography (possibly fault controlled), must have been subaerial and subject to strong erosion: From this was derived the molasse-like conglomerate facies. The clasts were most probably deposited in channels and fans; their well-rounded nature is probably a consequence of fluvial transport and also because their mafic composition made them particularly likely to wear uniformly in friction rather than to fracture in collision with other clasts.

The shallow-water and intertidal argillites may represent the characteristic deposit away from alluvial fans. Carbonate deposition must have taken place in specially protected settings, however, and most probably occurred in lagoons well protected both from long-shore drift and from the influx of terrigenous debris from streams. Martin (1983) has suggested that the western subsidiary syncline in which most limestones occur represents a discrete depository. The evidence for or against this hypothesis is ambiguous, but it is clear that the region of carbonate deposition must have been quite protected and discrete. Near Mumburumu peak it is interesting that a very thick conglomerate lens is juxtaposed to a carbonate mass: Most probably the conglomerate was laid down on a fan in a topographic low and the carbonate reef grew in shallow water protected by barriers and topography.

Figure 6.39. Outcrop of conglomerate, Cheshire Formation, Zeederbergs Siding.

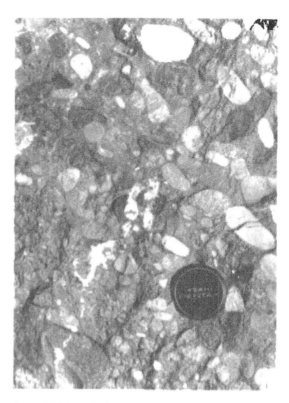

Figure 6.40. Detail of conglomerate, Zeederbergs Siding.

Figure 6.41. Stromatolites, limestone member, Cheshire Formation. Scale given by hammer head left bottom.

160

6.8 DIRECTIONS OF CURRENT MOVEMENT AND THE NATURE OF THE DEPOSITIONAL BASIN

Many Archaean greenstone belts have been described as 'basins' (see Walker 1978) on the basis of very little evidence. It is not obvious from the lateral facies variations whether the rocks of the Belingwe Belt as now preserved were deposited in isolated sedimentary basins centred on their present outcrop area or whether greenstone belts preserve parts of a more widespread sedimentary and volcanic sequence. Palaeocurrent analysis may permit the distinction between these two sedimentary models. Obviously palaeocurrent data should not be considered in isolation as sedimentary facies variations should also mirror the geometry of the depositional basin. Limitations in outcrop pattern may obscure facies variation, particularly if the present synclinal structure was centred on the original depositional basin. Studies of modern and recent sedimentary basins and of sedimentary processes all emphasise the possible complexities in the analysis of palaeo-current indicators, particularly hierarchical arrangements of several current directions. However palaeocurrents derived from cross-bedded sediments in general show transport towards the deeper parts of their sedimentary basins.

The sophistication of the measurements described here is severely limited by the complexity of the unfolding necessary and the limited number of localities measured in any stratigraphic horizon. The overall consistency of the observed current directions, towards the north-west, is taken to imply that sediment transport was consistent in direction throughout the deposition of the greenstone belt and that the measured directions are significant.

6.8.1 *Measurement and correction of current directions*

Current directions are derived from the dip of cross-stratified foresets in sediments with cross-bedded units between 5 and 30 cm thick, as well as from ripple orientations in ripple drift cross-laminated sediments. In some ripple-drift cross-laminated sediments only long axes of ripples could be measured with no single transport direction apparent. These data are recorded on two-way rose diagrams (Fig. 6.45).

All the measurements require several rotations to correct for the multi-phase folding of the belt. In addition internal strain has been ignored. Internal strain throughout much of the Belingwe Greenstone Belt is less than 30% (Coward et al. 1976) and at localities where cross-bedding is preserved it is much less than this. The maximum strain of 30% would rotate poles at most by 10° which is insignificant. Correction by simple rotations about fold axes is valid only if the folding mechanism is flexural slip (Ramsay 1967). The observation that deformation, particularly during the tight to isoclinal D_g2 phase, was restricted to narrow bedding plane shears with the intervening sediment remaining undeformed is consistent with this fold mechanism. In this case most of the error in the unfolding processes will be due to uncertainty in the plunge of the fold axes. The observed current directions were corrected as follows:

First the rotations due to D_g3 and D_g4 folding, about steeply plunging axes, were eliminated by one rotation about a vertical axis such that the D_g2 axis was rotated to an orientation of about 340°. Secondly the D_g2 axis was rotated to the horizontal and finally beds were rotated to the horizontal about the D_g2 axis. It is the plunge of the D_g2 axis that is least well defined. Where the D_g2 closure is mapped in the southern part of Cheshire it plunges 85° north. Elsewhere D_g2 folds are rarely observed. The subsidiary D_g2 syncline through Mumburumu to the main stromatolite locality has a nearly horizontal plunge. Small folds exposed in the Ngezi in the east of Cheshire also have subhorizontal plunges.

Figure 6.42. Detail of stromatolite morphology, main outcrop. Each bed has a distinctive characteristic morphology (Martin et al. 1980).

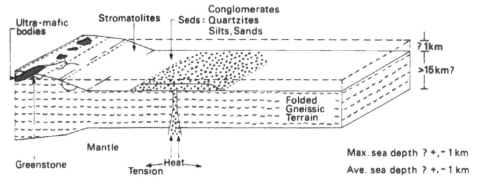

Figure 6.43. Paleogeographic reconstruction of local conditions at onset of Reliance-Zeederbergs volcanism.

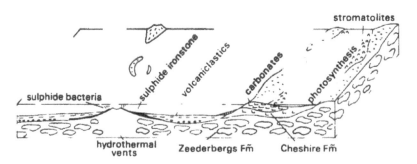

162

Figure 6.44. Paleogeographic reconstruction of local conditions at onset of Cheshire sedimentation.

Similarly small folds in the limestone band south of Mumburumu are subhorizontal. For the localities in the centre of the belt the assumption that D_g2 is horizontal is consistent with the available evidence. At locality 6, D_g2 plunges are between 10 and 40° northeast and accurate correction is possible. Locality 7 is only 1 km north of the nearly vertically plunging D_g2 closure although the small subhorizontal D_g2 folds occur 2 km west of this locality. The assumption of a horizontal axis may thus result in an error of up to 90° for this locality. If the D_g2 axis did plunge vertically at this locality the corrected mean current direction would be 90° west of that displayed in Figure 6.45. Similarly if at locality 1 the D_g2 plunge were towards the south the actual current direction might also be up to 90° west of that calculated. A similar correction procedure was used for the two localities in the Brooklands Formation.

6.8.2 *Interpretation of the results*

The results are presented in Figure 6.45 and in Table 6.1 which includes data from all formations. Where both planar cross-bedding and ripple-drift cross-lamination occur at the same locality no significant difference in orientation is discernible. Similarly two-way rose diagrams of current poles from symmetrical ripples are not significantly different from the directional data at the same localities. Most of the localities show well developed maxima towards the north or north-west. The Mtshingwe River reference section of the Manjeri Formation with 'herring-bone cross-bedding' shows principal directions towards the north-west, south-east and south-west. Even if the Manjeri type section localities and the southern Cheshire locality should be plotted with more westerly orientations to correct for errors in D_g2 plunge estimates, the seven localities from the main greenstones show surprising uniformity in indicating currents towards the north-west. There is no evidence that the D_g2 syncline was developed on the site of an earlier sedimentary basin with sediments supplied from elevated areas to the east and the west, although it should be noted that the main transport direction is sub-parallel to the D_g2 fold axis. It is surprising and perhaps instructive that the transport direction remained un-

Table 6.1. Current directions restored to horizontal.

Location*	No. of observations	Mean	1σ
1. Manjeri type locality			
Structure of definite direction	20	326	14
Bi-directional structures	21	323	24
2. Mtshingwe River outcrop of Manjeri Formation			
Cross-bedding (270-30°)	5	332	21
Cross-bedding (30-270°)	8	158	47
3. Mtshingwe Quarry in Manjeri Formation			
Directional structures	8	345	5
Bidirectional structures	6	344	12
4. Zeederbergs tuffs	9	338	17
5. Cheshire sandstones	14	308	32
6. Mumburumu sediments, Cheshire Formation	9	350	70
7. Southern Cheshire Formation			
Directional structures	14	351	26
Bi-directional structures	1	144	–

*See Figure 6.45.

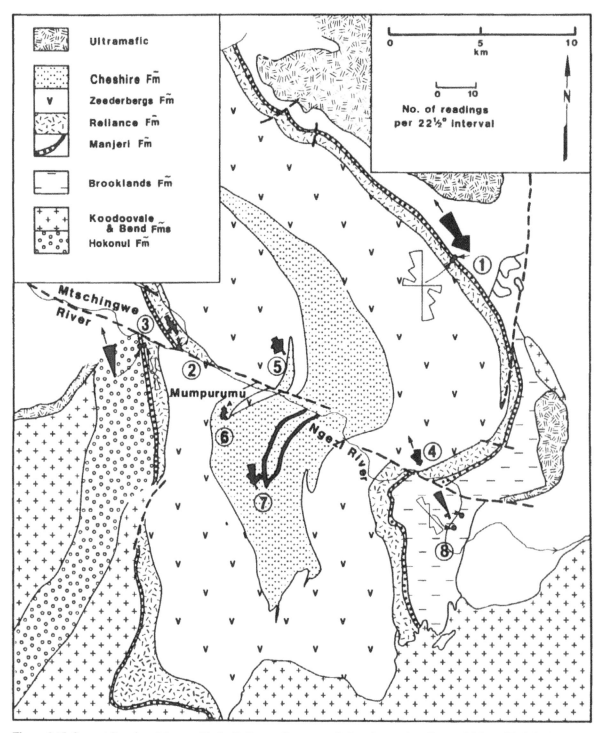

Figure 6.45. Current directions. Measured in the Belingwe Greenstone Belt and rotated to allow unfolding of D_g2, D_g3 and D_g4 deformations. Plotted relative to D_g2 fold axis trending at 340°.

changed during the deposition of the complete greenstone sequence. The single locality in cross-bedded tuffs in the Zeederbergs Formation is consistent with the volcanic formations being deposited in a similar topographic environment. Similarly, although the two localities in Brooklands are too few to yield a definitive direction for that Formation, the measured currents also trend to the north-west. However the consistency of the directions at these localities suggests that further measurements both in Belingwe as well as in the surrounding greenstone belts might yield useful information.

CHAPTER 7

Pb and Nd isotopic correlation in Belingwe komatiites and basalts

C. CHAUVEL, B. DUPRE & N.T. ARNDT

ABSTRACT

Komatiites and basalts from two locations in the Ngezi Group of the Belingwe Green-stone Belt define parallel Pb-Pb isochrons with ages of 2692± 9 Ma and 2675±173 Ma, and μ_1 values of 8.2 and 8.4. Data from a third location scatter and do not give useful age information. The Pb-Pb ages are consistent with reliable ages from other greenstone belts in Zimbabwe, and are believed to be correct. Nd isotopic data on the same samples form a linear array in the Sm-Nd diagram with a slope corresponding to about 2.9 Ga. Initial ε_{Nd} values, calculated at 2.7 Ga, range from +3.2, to + 0.4 and show a negative correlation with the μ_1 values. The anomalously old Nd age and the co-variation in initial ε_{Nd} and μ_1 values are explained by contamination of the lavas by old surrounding basement.

7.1 INTRODUCTION

The ages of Archaean mafic and ultramafic volcanic rocks, particularly those affected by metamorphism and hydrothermal alteration, are not easily determined. Numerous inves-tigations in the past 20 years have shown that the Sr system is commonly reset in such rocks, and more recent work has demonstrated that Sm-Nd ages are unreliable because of variations in initial isotopic ratios introduced by crustal contamination (Cattell et al. 1984; Chauvel et al. 1985). Under such circumstances the Pb-Pb whole rock approach becomes a valid alternative. In this paper we present the results of an investigation of Sm-Nd and Pb-Pb isotopes in samples of mafic and ultramafic rocks from three parts of the Belingwe Belt. In this case, the Pb-Pb method gave what appears to be an accurate age whereas the Sm-Nd method produced a result that is clearly too old.

7.2 SAMPLE DESCRIPTIONS

The three sections sampled in this study are shown in Figure 7.1. Section A, which is 5.3 km south of Zvishavane, is the source of the unusually fresh komatiites described by Nisbet et al. (1977) and Renner et al. (1991). At this locality, two spinifex-textured, layered komatiite flows were sampled. The lower flow provided samples Z2, Z3 and Z5, and the overlying flow, samples Z6, Z8 and Z9. The flows are shown diagramatically in

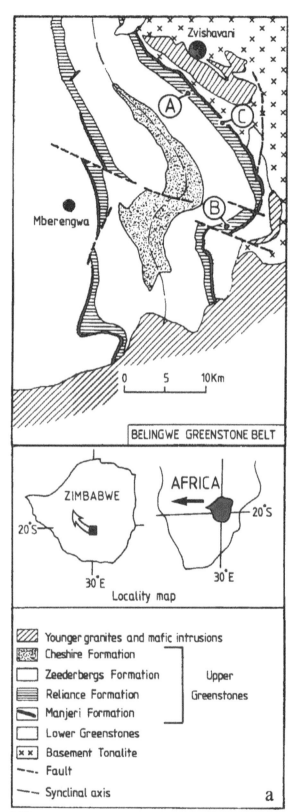

BELINGWE GREENSTONE BELT

ZIMBABWE

AFRICA

20°S

20°S

30°E

30°E

Locality map

- ▨ Younger granites and mafic intrusions
- ▨ Cheshire Formation
- ☐ Zeederbergs Formation ⎫ Upper
- ▤ Reliance Formation ⎬ Greenstones
- ◣ Manjeri Formation ⎭
- ☐ Lower Greenstones
- ⊠ Basement Tonalite
- ∙—∙ Fault
- — — Synclinal axis

a

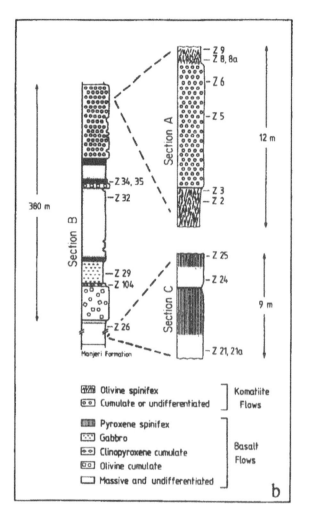

380 m

Section B

Section A

12 m

Z 9
Z 8, 8a
Z 6
Z 5

Z 3
Z 2

Z 34, 35
Z 32

Z 29
Z 104

Z 26

Manjeri Formation

Section C

9 m

Z 25
Z 24

Z 21, 21a

- ▨ Olivine spinifex ⎫ Komatiite
- ⊙⊙ Cumulate or undifferentiated ⎬ Flows

- ▥ Pyroxene spinifex ⎫
- ☐ Gabbro ⎪
- ⊙⊙ Clinopyroxene cumulate ⎬ Basalt
- ⊙⊙ Olivine cumulate ⎪ Flows
- ☐ Massive and undifferentiated ⎭

b

Figure 7.1. a) Simplified map of the Belingwe Greenstone Belt in Zimbabwe. New names for the towns are: Zvishavane = Shabani and Mberengwa = Belingwe. The three sections A, B, C are in the Reliance Formation and are shown by circles. b) Simplified sections of sections A, B and C showing their relative stratigraphic position as well as the samples locations.

Table 7.1. Pb isotopic data on Belingwe Upper Greenstone Belt.

Sample	$^{206}Pb/^{204}Pb$	$^{207}Pb/^{204}Pb$	$^{208}Pb/^{204}Pb$	μ_1
Location A				
Z3	18.42	15.62	37.33	
Z5	18.38	15.61	37.30	
Z6	18.76	15.68	37.60	8.2
Z8	18.41	15.61	37.30	
Z9	18.55	15.63	37.43	
Location B				
Z26	16.88	15.58	36.43	
Z32	52.61	20.13	74.97	
Z38B	19.86	16.03	38.94	
Z35	41.26	19.17	61.57	
Location C				
Z21	17.32	15.50	36.87	
Z21*	17.16	15.47	36.75	
Z21a	21.14	16.20	40.09	
Z21a*	18.81	15.77	37.88	8.4
Z24	22.51	16.45	41.81	
Z24*	21.91	16.34	41.15	
Z25	16.60	15.36	36.18	
Z25*	15.58	15.17	35.15	

*: Analysis of a different chip from the sample. The errors on the measured ratios are: $^{206}Pb/^{204}Pb$ and $^{207}Pb/^{204}Pb$:0.05%, $^{208}Pb/^{204}Pb$:0.10%. The apparant μ values (μ_1) have been calculated using isochrons for samples from sections A and C, and an age of 4.55 Ga for the Earth.

Figure 7.2. Brief sample descriptions are given in Table 7.1 and accounts of the petrography and chemistry are to be found in previous chapters.

Section B is the type-section of the Reliance Formation on the Lou Estate (Nisbet et al. 1977, 1987). Here the samples came from four different basaltic flows, one tholeiitic and the other three komatiitic (Fig.7.2). Section C lies just above the type-section of the Manjeri Formation in the northeast of the belt. This section provided four samples from two layered, pyroxene spinifex-textured basaltic flows. Samples from sections B and C are more highly metamorphosed than those from section A, and have sub-greenschist facies mineral assemblages. Textures are well preserved but primary minerals are partially to completely replaced by secondary assemblages.

7.3 PREVIOUS ISOTOPIC STUDIES OF THE BELINGWE GREENSTONES AND OTHER ARCHAEAN UNITS OF THE ZIMBABWE CRATON

Extensive geochronological studies have been carried out on the greenstone belts, and surrounding granite-gneissic terrains of the Rhodesian Craton in Zimbabwe (see Chapter 3). These have been summarized by Wilson (1979) and more recently by Taylor et al. (1991), who also supplied new Pb-Pb ages. According to these authors, the craton consists of 1) a small area of ca. 3.5 Ga old felsic gneisses and supracrustal rocks (the Sebakwian Group); 2) more extensive 3.0-2.9 Ga old granitoids, gneisses and volcanic rocks of the lower Bulawayan Group or lower greenstones; 3) widespread 2.7-2.6 Ga old volcanic and sedimentary rocks of the upper Bulawayan Group or upper greenstones;

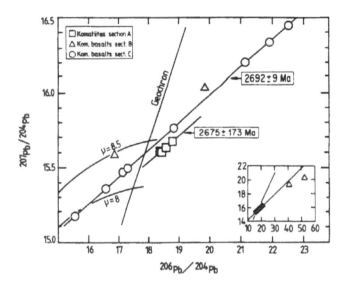

Figure 7.2. $^{207}Pb/^{204}Pb$ vs. $^{206}Pb/^{204}Pb$ diagram showing the two parallel isochrons. The isochrons ages are shown in boxes.

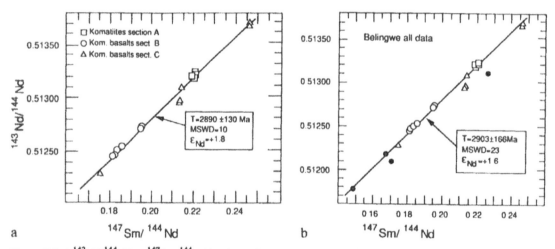

Figure 7.3. a) $^{143}Nd/^{144}Nd$ vs. $^{147}Sm/^{144}Nd$ isochron diagram showing how the data points scatter about the 'isochron'; b) Same diagram showing the relative positions of our samples compared to data published by Hamilton et al. (1977) shown by black dots.

and 4) ca. 2.6 Ga granitoid pluton which intrude all early rocks. Although the majority of ages are from Rb/Sr whole-rock studies, e.g. (Hawkesworth 1975, 1979; Moorbath 1976, 1977; Jahn & Condie 1976; Taylor et al. 1991), certain units have also been studied by Sm-Nd whole-rock (Hamilton et al. 1977; Moorbath et al. 1986) and Pb-Pb whole rock methods (Taylor et al. 1984, 1991; Moorbath et al. 1987).

Previous geochronological studies of the 2.7 Ga old upper Greenstones, of which the Belingwe Belt is a member, included a Rb-Sr investigation by Hawkesworth et al. (1975) who obtained an isochron age of 2485 ± 90 Ma. Jahn & Condie (1976), by combining their whole rock analyses with those of Hawkesworth et al. (1975), obtained a Rb-Sr isochron age of 2700 ± 70 Ma for the Zeederbergs volcanics in the Ngezi Groups. Hamilton et al. (1977) combined data from samples of four different belts within the Rhodesian craton to obtain a Sm/Nd isochron. Their data (including four volcanics from the Reliance Formation in the Ngezi Group) define an age of 2640 ± 140 Ma and an initial ε_{Nd} value of ~ 0. Although the samples come from different, widely separated belts, they interpret this age to reflect the extrusion time and the initial ε_{Nd} value to be representative of the mantle source.

7.4 RESULTS

Analytical techniques for both Pb and Nd isotopic measurements have been described elsewhere (Chauvel et al. 1985; Dupré & Echeverria 1984; White et al. 1985). Pb results are presented in Table 7.1 and Figure 7.2, and Nd data in Table 7.2 and Figure 7.3.

Komatiitic basalts from section C, the region of komatiitic basalts close to the type-section of the Manjeri Formation, define a Pb-Pb isochron with an age of 2692 ± 9 Ma (Fig.7.2). The apparent μ_1 value (calculated assuming 4.55 Ga for the age of the Earth) is 8.4. In the $^{208}Pb/^{204}Pb$ versus $^{206}Pb/^{204}Pb$ diagram (not shown), these samples define a regression line which gives a Th/U ratio of 3.4 ± 0.1. The well-preserved komatiites from section A have only a small range of $^{206}Pb/^{204}Pb$ ratios but are nevertheless collinear and give an age of 2675 ± 173 Ma. This second isochron is almost parallel to the first, but lies distinctly below it (Fig. 7.2). The apparent μ_1 value of 8.2 is distinctly lower than that

Table 7.2. Nd isotopic results for the Belingwe Upper Greenstone Belt.

Sample	Nd (ppm)**	Sm (ppm)**	$^{147}Sm/^{144}Nd$**	$^{143}Nd/^{144}Nd$	$\pm 2\sigma$	ε_{Nd}
Location A						
Z2	2.41	0.882	0.2212	0.513197	18	+2.36
Z5	1.85	0.674	0.2197	0.513189	25	+2.72
Z6	1.73	0.628	0.2203	0.513177	26	+2.28
Z9	2.12	0.776	0.2210	0.513222	16	+2.91
Location B						
Z32	4.33	1.25	0.1756	0.512284	15	+0.39
Z34b	2.76	0.975	0.2136	0.512963	30	+0.43
Z34b*	2.75	0.97	0.2133	0.512952	28	+0.32
Z29	1.71	0.608	0.2144	0.513078	19	+2.40
Z104	0.97	0.396	0.2463	0.513647	40	+2.41
Z104*	0.97	0.396	0.2466	0.513693	23	+3.2
Location C						
Z21	4.21	1.28	0.1836	0.512511	18	+2.04
Z21a	5.19	1.59	0.1859	0.512540	24	+1.81
Z24	4.05	1.22	0.1821	0.512473	15	+1.82
Z24*	4.22	1.27	0.1819	0.512469	20	+1.81
Z25	4.50	1.46	0.1956	0.512732	40	+2.18
Z25*	4.59	1.48	0.1957	0.512728	22	+2.07

*Duplicate analysis. **Error on the ratio $^{147}Sm/^{144}Nd$ is 0.4%. The error on the absolute concentrations of Nd and Sm is far larger due to the aliquoting technique. $\varepsilon_{Nd}(T)$ are calculated for an age of 2.7 Ga as deduced from the Pb-Pb isochrons.

calculated for samples from section C. Samples from section B scatter and do not define an isochron.

In the Nd isochron diagram (Fig. 7.3), samples from all sections plot in a linear array with a slope corresponding to an age of 2890 ± 130 Ma. This age is significantly different from that obtained by Hamilton et al. (1977) (2640 ± 140 Ma) but, in both cases, the alignment of the data points is very poor, as indicated by the large errors on the ages and the high MSWD values. Attempts to define separate isochrons for each of the three sections were unsucccessful because of the small variation of Sm/Nd ratios in rocks from section A ($^{147}Sm/^{144}Nd = 0.22$ to 0.23) and section C ($^{147}Sm/^{144}Nd = 0.18$ to 0.20). Samples from section B show a wider spread in Sm/Nd but are not collinear.

7.4.1 *Age of the Ngezi Group, Belingwe Greenstone Belt*

The ages obtained in this study show a pattern that is becoming familiar in investigations of Precambrian volcanic rocks: The Sm-Nd age is significantly greater than the Pb-Pb age. We suggest that the latter age is more likely correct, for the following reasons. First, the Pb data define a good isochron for the komatiitic basalts from section C with an error of only 9 Ma. Pb isotopes have been shown to be a reliable dating method in the case of Precambrian komatiites and basalts from many other regions (see summary in Dupré & Arndt 1986) in which Pb-Pb ages are indistinguishable from U-Pb zircon or other reliable ages. In contrast, it has now been demonstrated that the Sm-Nd method commonly gives ages that are too old e.g. Newton, Ontario, Cattell et al. (1984); Usushwana complex Hegner et al. (1984); Kambalda, Western Australia Chauvel et al. (1985); Compston et al. (1986). We therefore propose that the Pb-Pb isochron from the basalts of section C gives a reliable age for the Ngezi Group and that these rocks are 2692 ± 9 Ma old.

7.4.2 *Primary Nd and Pb heterogeneities*

Since samples from sections A and C define parallel Pb-Pb isochrons, they should have had different initial isotopic ratios at 2700 Ma. From the isotopic data we calculate the following model μ_1 values ($^{238}U/^{204}Pb$) for the Pb in the sources of the two sets of komatiitic lavas: 8.2 for the komatiites of section A, and 8.4 for the komatiitic basalts from section C.

The initial ε_{Nd} value for each sample was calculated using the 2.7 Ga Pb-Pb age (Table 7.2). All samples have positive values, with an average of $+2.6 \pm 0.4$) and a total range from $+0.3$ to $+3.2$. The well-preserved komatiites from section A have a restricted range of ε_{Nd} values, from 2.3 to 2.9. These values are commensurate with the higher-than-chondritic Sm/Nd ratios of the rocks ($^{147}Sm/^{144}Nd = 0.22$) and indicate that they came from a source depleted both chemically and isotopically. The basalts from section C have only slightly lower ε_{Nd} values, from 1.8 to 2.2, but their $^{147}Sm/^{144}Nd$ values range from chondritic to slightly enriched (0.196-0.182). Although their source was isotopically depleted, its chemical composition must have changed slightly before the time of eruption, or a fractionation process such as low-degree partial melting, or crystallization $\pm$ contamination must have affected the compositions of the magmas.

Komatiitic basalts from section B are uniform neither in Sm/Nd ratios (0.17-0.21) nor in initial ε_{Nd} values ($+0.3$ to $+3.2$). These lavas have been sampled from sites widely dispersed within the 400 m section of tholeiitic and komatiitic basalts (see Fig. 7.1b). The variations in Sm/Nd ratios and ε_{Nd} values can be attributed to variations of trace element chemistry and isotopic compositions of different lava flows, perhaps augmented by the effects of alteration of these rocks, which are not as well preserved as those from the other two sections.

From the data listed in Tables 7.1 and 7.2 and plotted in Figures 7.2 and 7.3, it can be seen that there is a simple negative correlation between ε_{Nd} and μ_1. The komatiites from section A have high ε_{Nd} (+2.6) and low μ_1 (8.2); the basalts from section C have lower ε_{Nd} (+2.0) but higher μ_1 (8.4). Komatiitic basalts from section B have variable ε_{Nd} values but, since they do not define a Pb-Pb isochron, no correlation with μ_1 can be made.

7.5 ORIGIN OF THE HETEROGENEITIES

We suggest that the variations in μ_1 and ε_{Nd} values obtained for the volcanics in the Belingwe Ngezi Group are best explained by assimilation of older felsic crustal material by the lavas. This interpretation is consistent with the well publicized propensity of komatiites to thermally erode and assimilate whatever rock they encounter during their eruption (Huppert et al. 1984; Huppert & Sparks 1985) and with the presence in the area of old granitoid basement (~ 3.5 Ga) underlying the volcanics.

A reasonable candidate for the older contaminant is the Shabani Gneiss which is ca. 3.5 Ga old, according to Rb/Sr dating of Moorbath et al. (1977) and outcrops northeast of the Belingwe Belt. Pb-Pb dating of these rocks by Taylor et al. (1991) gave 3.0 Ga and a high μ_1 of 9.0, a result that is attributed by Taylor et al. to U-gain during a later event that reset the Pb system. As shown in Figure 7.4, less than 0.5% of contamination of the komatiitic liquid by such a continental material accounts for the Nd and Pb isotopic compositions

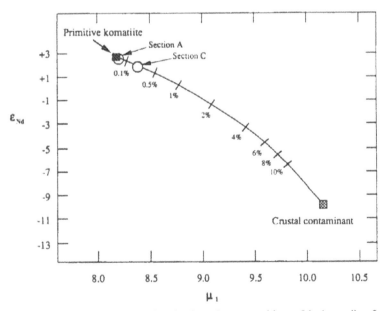

Figure 7.4. ε_{Nd} vs. μ_1 diagram showing how the compositions of the komatiites from sections A and C can be explained by crustal contamination at 2.7 Ga. The composition of the contaminant is based on data from the Shabani Gneiss whose μ_1 can be estimated at 10.2 using the results of Taylor et al. (1991). If the Nd isotopic composition was chondritic at 3.5 Ga, as suggested by the model age data of Moorbath et al. (1986), its ε_{Nd} (t = 2.7Ga) would have been -10 to -12. Pb and Nd concentrations are estimated as 8 and 32 ppm, respectively. The komatiite end-member has μ_1 = 8.2, ε_{Nd} = +3, 0.2 ppm Pb and 1.5 ppm Nd. Tick marks on the mixing curve show the percentage of crustal component.

173

measured on the komatiitic basalts from section C, and about 1% contamination is required to explain the ε_{Nd} values close to 0 measured on the komatiitic basalts from section B.

Contamination with older felsic rocks may also have affected many of the samples analysed by Hamilton et al. during their Sm-Nd study. Their age of 2640±140 Ma is correct within its large error, but the initial ε_{Nd} calculated from this isochron is ~ −1, significantly lower than the value for depleted mantle that we infer from the compositions of the least-contaminated Belingwe komatiites, and lower than values inferred during other studies of ~ 2.7 Ga mafic rocks (e.g. Chauvel et al. 1985; Shirey & Hanson 1986). We suggest that the Hamilton et al. age probably is a fortuitous result due to the presence in the sample suite of low Sm/Nd rocks that are not significantly more contaminated than the samples with high Sm/Nd.

A different process must be called on to explain the Pb isotopic compositions of the two basaltic samples from section B that have extremely radiogenic compositions and lie well below the isochrons. These samples are located close to a major fault and to a late granitic intrusion (Fig. 7.1a) and their Pb systematics could have been disturbed both during the granitic intrusion and/or during the faulting event. These two processes could have involved an exchange of U (and Pb) between the mafic volcanics and the surrounding material, and addition of some U to the original composition of the lavas.

ACKNOWLEDGEMENTS

We thank Tony Martin for his help during sample collection. B.D. thanks the Max Planck-Foundation for support during his stay in Mainz. N.T.A. and C.C. received support from the Deutsche Forschungsgemeinschaft projects Kr 590/11.1 and Kr 590/11.2. Steve Moorbath and Euan Nisbet provided useful comments on a version of the manuscript whose age approaches that of the rocks we studied.

CHAPTER 8

Geochemistry of the igneous rocks of the Belingwe Greenstone Belt: Alteration, contamination and petrogenesis

M.J. BICKLE, N.T ARNDT, E.G. NISBET, J.L. ORPEN, A. MARTIN, R.R. KEAYS & R. RENNER

ABSTRACT

The Belingwe Belt contains an andesitic suite, two suites of komatiites associated with voluminous tholeiites, and assorted intrusive igneous rocks. Some of the latter include remarkably fresh komatiite and basalt, containing little altered olivine and clinopyroxene phenocrysts and optically fresh glass. However most rocks are altered, probably both by early hydrothermal and later regional metamorphism, as well as by recent weathering. Within the very fresh flows in the Reliance Formation SiO_2, TiO_2, Al_2O_3, FeO, MgO, Na_2O, NiO, Cr_2O_3, Zr, Y, REE and platinum group elements vary in a manner consistent with models which involve fractionation of olivine of the observed compositions and consistent with experimentally determined partition coefficients. These elements were probably nearly immobile during alteration.

The Bend and Reliance komatiite suites are compositionally distinct in incompatible trace element ratios (e.g. Zr/Ti, Y/Ti). Crustal contamination of komatiite lavas is probably less than 1% by mass. The lowermost volcanic unit, the Hokonui Formation is, in contrast, a typical calc-alkaline sequence.

8.1 INTRODUCTION

The Belingwe Greenstone Belt contains several sequences of extrusive Archaean igneous rocks, including an andesitic suite and two suites of komatiites and tholeiites, in addition to intrusive igneous rocks. Although all the rocks have been mineralogically altered to some extent, some specimens are as little altered as many Tertiary lavas. The best samples from Belingwe rank amongst the freshest of all Archaean lavas. The geochemistry of the volcanic rocks of the Ngezi Group has been extensively discussed elsewhere (Nisbet et al. 1977; Bickle et al. 1975, 1977; Cameron & Nisbet 1982; Chauvel et al. 1983; Nisbet et al. 1982, 1987; Renner et al., in press). We present here further geochemical data on komatiitic and basaltic rocks from the Ngezi Group and new major and trace element data from the komatiitic, basaltic and calc-alkaline lavas of the Mtshingwe Group.

Archaean komatiitic and calc-alkaline igneous suites constrain global thermal and geochemical evolution. The high magnesium komatiitic lavas, which represent high degrees of mantle melting, place constraints on the incompatible element concentrations

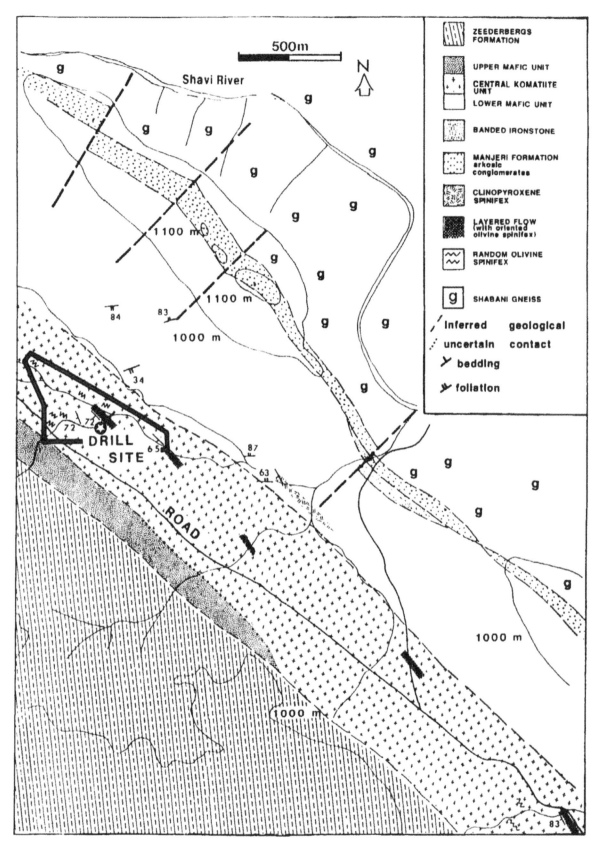

Figure 8.1. Geological map of the vicinity of the SASKMAR 1 site. Geology of drill site (outlined with heavy line) shown in Figure 8.2.

of the source regions. Komatiites also provide evidence for higher Archaean mantle temperatures (e.g. Bickle 1986), predicted by models of global thermal evolution (Richter 1985; Christansen 1985). However the physical mechanisms responsible for the generation of komatiites are still problematic, as the generation of melts with MgO and incompatible element contents typical of komatiites requires adiabatic upwelling from depths in excess of 400 km and enormous volumes of melt would have been produced (e.g. Bickle 1986; McKenzie & O'Nions 1991). Calc-alkaline igneous activity is probably a significant mechanism of crustal growth. Secular changes in crustal and mantle composition are predicted by the simpler models for global geochemical evolution (Jacobsen & Wasserburg 1979; O'Nions et al. 1979). However Archaean calc-alkaline rocks exhibit surprising geochemical similarities to their modern counterparts (e.g. Bickle et al. 1983b; in press).

Most of the Archaean komatiitic volcanic rocks that have been described in the literature have been extensively altered, with primary igneous minerals being completely or mostly replaced by metamorphic assemblages. The possible geochemical consequences of alteration have raised doubts as to the precise significance of their geochemistry and particularly their MgO contents and eruption temperatures (e.g. Elthon 1986; de Wit & Hart 1986). Furthermore Huppert et al. (1984) and Huppert & Sparks (1985a, b) concluded that such high-temperature magmas would have melted their wall-rocks during transport through the crust or thermally eroded the ground during eruption. Here we present evidence from the petrology, mineralogy and chemistry of some very fresh layered komatiitic flows in the Ngezi Group that geochemical alteration has not significantly affected the komatiite chemistry. The new trace element data, in conjunction with the Nd and Pb isotopic data discussed in Chapter 7, also puts limits on the extent of crustal contamination during eruption. Analyses of the previously unpublished data from komatiitic lavas within the Mtshingwe Group are compared with the stratigraphically distinct but spatially juxtaposed komatiite-basalt sequence of the Ngezi Group.

8.2 SAMPLE SUITES AND SELECTION

Previously analysed samples from the Ngezi Group were described by Bickle et al. (1975) and Nisbet et al. (1977, 1987) and were mainly collected from the type section on the Lou Estate. Some other Ngezi Group samples were collected from the localities described by Martin (1978) on the western side of the belt. Analyses of some very fresh samples, collected from a locality in the Reliance Formation 5.3 km south of Zvishavane (Fig. 8.1, GR 906466), are reported here. This relatively poorly exposed locality has subsequently been drilled to a downhole depth of 200 m (SASKMAR 1 hole, Nisbet et al. 1987) and the core provides a continuous 170 m section through a sequence of thin, massive or layered spinifex-textured flows which may be correlated with the less completely exposed flows exposed at the surface (e.g. Fig. 8.2b).

The Mtshingwe Group samples which have been analysed include a suite from the Bend and Koodoovale Formations (Orpen 1978), the Hokonui Formation (Martin 1978; Orpen 1978) and the Brooklands Formation. Localities are given in the appendix. Almost all samples are from large specimens hammered off surface outcrop and although every effort was made to collect the freshest material in clean outcrop it was in general impossible to avoid some surface weathering as discussed below. Samples sets NG, BL and B were collected by Bickle and Nisbet; AM by Martin; SL and GC by Orpen; RL by Bickle, Hawkesworth and Nisbet; Z by Arndt and ZV by Renner, Cheadle, Bickle and Nisbet.

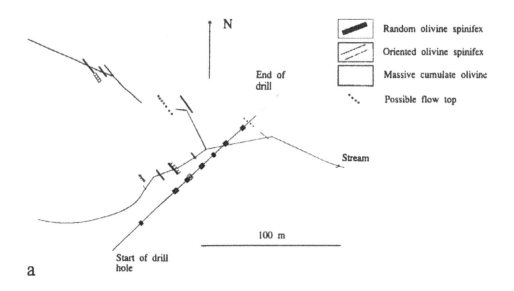

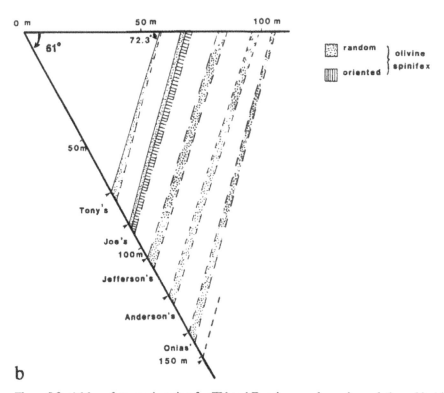

Figure 8.2. a) Map of exposed section for ZV and Z-series samples and correlation with drill core projected to surface. b) Correlation between surface and drill core. Assumes surface dip of lavas (72.3°) is constant with depth and dip of drill hole is constant at 60° measured at surface and 100 m downhole. Compiled by Renner (1989).

178

Table 8.1. Nb-Zr-Y-Sr-Rb values for USGS standards.

	Nb	Zr	Y	Sr	Rb
1. Values adopted as standards					
GSP-1	25.4	543	25.4	235	252
BCR-1	11.3	192	33.4	328	46.7
2. Values measured on USGS standards with above as standards, Leeds calibration					
AGV-1	13.4	241	18.7	665	67
G2	11.2	323	9.9	485	168
3. Values measured on USGS standards, ETH calibration					
GSP-1	27.4	547	27.8	237	254
AGV-1	11.5	240	17.8	656	69
G2	7.0	321	9.6	477	166
W1	11.7	103	24.0	193	27

8.3 ANALYTICAL METHODS

New major element analyses of sample series SL, AM and NG (Table 8.7) from the Hokonui, Bend, Koodoovale and Brooklands Formations are by standard XRF methods at ETH-Zurich, using duplicate fused discs with a lithium tetraborate flux and no heavy absorber with careful drift control. Correction was by an alpha-factor matrix, iterated for inter-element effects. Calibration was with USGS and NIM Standards. Nb, Zr, Y, Rb and Sr analyses at Leeds were in triplicate by XRF on pressed powder pellets made with a moviol base with count times up to 1000 s. Calibration was based on USGS samples GSP-1 and BCR-1. Adopted values for these standards are given in Table 8.1 along with measured values for USGS standards AGV-1 and G-2. Interference corrections Y on Nb, Sr on Zr and Rb on Y (about 5, 10, and 30% of the interfering peak respectively) were measured on spiked blank samples of olivine. Background counts were calculated as a function of background readings between peak positions by repeated measurements on the blank olivine sample and on USGS standard PCC-1 (with correction for elements present). Mass absorption coefficients were taken as proportional to the inverse square of the Ag scatter peak. Values were compared with calculated mass absorption coefficients from major element compositions and agreement was always better than 4%. Komatiites contain rather low concentrations of these trace elements and use of very long count times by repetitive cycling and use of an internal standard ensured reasonable precision. Duplicate analyses of all samples and triplicate analyses of some samples with independent interference, background and calibration measurements provide an estimate of the precision. Average deviation from the mean is 3% for elements in the range 10 to 20 ppm and 5% for elements in the range 4 to 10 ppm. The good correlation between Zr and MgO for the Reliance Formation peridotites (Fig. 8.9) suggests that precision better than ± 5% was obtained. Uncertainty in mass absorption coefficient (± 5%) and in the values adopted for the standards places limits on accuracy of the trace element measurements. Nb, Zr, Y, Rb and Sr analyses at ETH-Zurich followed the same method and were duplicated by the method of Nisbet et al. (1979) using 400 s count times and are less precise than the analyses at Leeds.

Chemical data for the Z and ZV-series rocks are listed in Table 8.3. The major element data on the Z-series samples are the averages of XRF analyses carried out at the Max Planck Institute using the technique of Palme & Jagoutz (1977), and at the University of

Table 8.2. Microprobe analyses of phases in Z-series komatiites.

	Olivines Sample Z2					Sample Z5 Large grain		Small grain	
	2040 RC	2040 RM	2020 SK	2023 SK	2024 SK	501/01 Centre	501/02 Margin	502/01 Centre	502/02 Margin
SiO$_2$	41.59	40.90	41.27	40.88	39.92	41.93	40.41	41.55	41.08
FeO	8.64	11.16	9.31	12.12	15.06	6.33	12.59	8.25	11.19
MnO	0.00	0.00	0.00	0.00	0.15	0.00	0.13	0.00	0.16
MgO	50.22	48.08	49.26	47.26	44.25	51.00	46.31	50.06	47.58
CaO	0.11	0.14	0.13	0.13	0.24	0.09	0.17	0.15	0.17
NiO	0.41	0.39	0.43	0.22	0.29	0.44	0.27	0.44	0.37
Cr$_2$O$_3$	–	–	0.10	–	0.09	–	–	0.15	–
Total	100.96	100.67	100.44	100.62	100.00	99.79	99.87	100.58	100.54
Fo%	91.2	88.5	90.4	87.4	84.0	93.5	86.8	91.5	88.3

	Sample Z6					Sample Z8				
	601-01 Centre	601-06 Large grain	601-07	601-09	601-13	604-14	8041 Centre	8043 Margin	8044 SK	8032 SK
SiO$_2$	41.83	41.75	41.80	42.12	41.12	41.14	41.58	41.08	40.85	39.85
FeO	6.71	6.67	6.82	6.67	8.99	11.51	7.59	9.25	10.29	15.30
MnO	?	0.00	0.00	0.00	0.00	0.27	0.00	0.16	0.00	0.15
MgO	51.27	50.82	50.86	50.90	49.97	48.01	50.39	49.00	48.19	44.39
CaO	0.09	0.10	0.14	0.06	0.07	0.08	0.10	0.13	0.11	0.15
NiO	–	0.48	0.38	0.51	0.28	0.51	0.37	0.40	0.28	0.17
Cr$_2$O$_3$	–	–	–	0.12	0.19	–	0.11	0.12	–	–
Total	100.30	100.00	100.00	100.38	100.62	101.51	100.14	100.13	97.72	100.00
Fo%	93.2	93.1	93.0	93.2	90.9	88.1	92.2	90.4	89.3	83.8

	Pyroxene analyses Sample Z2							
	2036 CPMM	2036 CPM	2038 CPCO	2039 CPCI	2310 CPMI	2312 CPCC	2313 CPCC	2314 OPCC
SiO$_2$	48.10	49.86	50.94	55.11	54.94	55.78	56.01	55.69
TiO$_2$	0.40	0.32	0.30	0.00	0.00	0.00	0.00	0.00
Al$_2$O$_3$	6.72	6.28	4.72	2.04	2.04	1.86	1.44	1.75
FeO	11.68	8.56	8.55	9.70	9.46	8.28	8.23	8.10
MnO	0.21	0.17	0.17	0.00	0.19	0.17	0.00	0.00
MgO	13.39	16.23	18.36	28.44	28.97	30.87	31.14	30.85
CaO	18.65	17.98	15.89	4.01	3.77	2.33	2.11	2.36
Cr$_2$O$_3$	–	0.53	0.60	0.61	0.69	0.81	0.61	0.84
Total	99.63	99.93	99.53	99.91	100.05	100.09	99.55	99.59
$\dfrac{Mg}{Mg+Fe}$	67.1	77.2	79.3	83.9	84.5	86.9	87.1	87.2

Table 8.2 (continued).

	Glass analyses Sample Z2					Z8	Chromites Sample Z2	
	2011	2013	2014	2016	2033	8063	2063	
SiO2	52.91	48.38	48.10	49.06	54.17	52.15	0.00	0.00
TiO2	0.46	0.27	0.61	0.37	0.52	0.51	0.42	0.62
Al2O3	15.36	10.15	17.20	18.58	17.72	16.22	12.34	14.50
FeO	12.66	10.44	14.56	12.22	9.66	12.17	16.21	23.90
MnO	0.13	0.00	0.21	0.13	0.00	0.17	0.00	0.29
MgO	5.74	18.36	7.55	4.22	3.12	6.39	14.81	13.20
CaO	9.09	4.59	6.61	9.58	9.12	7.33	0.00	0.18
Na2O	2.76	1.22	2.66	3.95	4.11	3.22	–	–
K2O	0.14	0.00	0.12	0.23	0.05	0.11	–	–
NiO	0.00	0.00	0.00	0.00	0.00	0.00	0.00	0.27
Cr2O3	0.00	0.00	0.00	0.00	0.00	0.00	59.49	49.94
Total	99.26	93.41	97.85	99.24	99.09	98.27	103.27	103.00

Melbourne using the technique of Norrish & Hutton (1969). ZV-series samples were analysed at ANU, Canberra, by W. Cameron using the Norrish & Hutton (1969) fusion technique. Trace elements were analyzed at Max Planck Institute, Ni and Cr on fused discs, the others on pressed pellets. Platinum group elements were analysed using the NAA technique of Keays et al. (1974), and REE were analysed by isotope dilution at MPI using the technique of White & Patchett (1984).

8.4 ALTERATION AND PETROGRAPHY

All the samples have undergone some mineralogical alteration. Olivine is commonly replaced entirely by serpentine and magnetite or chlorite in less magnesium-rich samples; clinopyroxene and chromite are characteristically little altered in the fresher samples selected for analysis; plagioclase is for the most part altered to albite, and the once glassy groundmass is variably devitrified and replaced by greenschist facies mineral assemblages. In the fresher Ngezi Group samples some relict olivine is preserved, particularly in the thicker flows and especially in the SASKMAR 1 locality discussed further below. In the less altered samples the groundmass has a glassy appearance but is always hydrated though to a varying degree. Rare unaltered calcic (An_{62}) plagioclase laths are found in rocks of basaltic composition (in the Bend Formation).

The timing of the mineralogical changes is difficult to constrain. As discussed in Chapter 3, the rocks are likely to have been subject to an early syn-volcanic hydrothermal alteration, syn-D_g3 greenschist facies metamorphism and recent weathering. The severity of the latter is illustrated by the complete serpentinisation of otherwise little altered ultramafic rocks both in the nearby Great Dyke and Shabani Ultramafic complex to depths of 100 m below the present surface.

The least altered samples (Z-series, Table 8.3) were collected from well preserved komatiite flows exposed in the bed of a small spruit on the eastern limb of the Ngezi Group, about 5.3 km south of Zvishavane. This location was the site drilled by the SASKMAR 1 hole (Nisbet et al. 1987). In this region (Fig. 8.1), although the area of outcrop is relatively restricted some rocks are very fresh indeed. This allows assessment of the geochemical changes associated with alteration. We first review in detail the

Table 8.3. Belingwe Komatiites SASKMAR 1 Locality. Major and REE analyses of Z-series samples. All data recalculated anhydrous with all iron as FeO. Z-samples analysed by N.T.A. in Mainz.

%	Z1	Z2	Z3	Z4	Z5	Z6	Z6-1	Z7	Z8	Z8A	Z8B	Z9	Z10	Z11
SiO_2	51.47	50.30	48.18	49.34	48.72	47.35	47.27	48.22	48.75	50.43	51.24	48.98	46.86	47.12
TiO_2	0.48	0.44	0.35	0.41	0.33	0.32	0.32	0.33	0.41	0.45	0.48	0.41	0.30	0.30
Al_2O_3	9.67	9.13	7.23	8.24	6.78	6.39	6.38	6.83	8.07	9.19	9.68	8.04	6.02	6.11
FeO^*	12.28	11.89	11.32	11.54	11.05	11.07	10.88	11.52	11.66	12.04	11.98	11.64	10.84	10.62
MnO	0.20	0.21	0.19	0.20	0.19	0.18	0.18	0.20	0.19	0.20	0.20	0.20	0.18	0.18
MgO	15.29	17.82	24.17	20.65	25.48	27.23	27.34	25.08	21.91	18.07	16.10	21.28	28.69	28.50
CaO	8.93	8.91	7.45	8.35	6.70	6.71	6.76	6.88	7.82	8.57	8.87	8.14	6.12	6.21
Na_2O	1.29	1.21	1.00	1.19	0.70	0.67	0.78	0.88	1.09	0.98	1.32	1.21	0.89	0.83
K_2O	0.07	0.05	0.06	0.07	0.03	0.03	0.04	0.04	0.06	0.05	0.11	0.07	0.06	0.09
P_2O_5	0.32	0.03	0.03	0.03	0.02	0.02	0.02	0.02	0.03	0.03	0.03	0.03	0.02	0.02
(H_2O)	2.30	2.30	2.90	2.70	4.00	3.20	2.30	3.30	3.00	4.20	2.70	2.90	4.20	4.40
ppm														
Rb	2.0	0.2			2.5	1.4		1.7	0.6	0.7	1.7	0.3	1.4	1.6
Sr	45	43	35	40	31	29	30	32	40	47	53	40	30	27
Ba	39	22	22	21	21	16	16	23	28	31	46	22	16	17
Zr	25	30	21	21	18	15	15	17	23	23	22	22	14	17
Sc	34	31	25	30	23	23	22	24	27	30	33	28	21	22
Cr	1735	2600	2720	2685	2470	2600	2500	3070	2680	2528	2270	2730	2630	2540
Ni	442		1150	844	1335	1410	1424	1310	945	606	486	930	1515	1500
La	1.15	1.02				0.801			0.809	1.05	1.01			0.606
Ce	2.85	2.69			2.00	1.87					3.41			1.65
Nd	2.73	2.46			1.90	1.81				2.51	2.74			1.58
Sm	0.990	0.899			0.691	0.665				0.896	0.980			0.590
Eu	0.397	0.364	0.265		0.279	0.266		0.300		0.354	0.400			0.240
Gd	1.45	1.34	0.982		1.05	0.973		1.09		1.35	1.45			0.890
Dy	1.82	1.68	1.22		1.30	1.25		1.35		1.68	1.82			1.12
Er	1.20	1.09	0.77		0.854	0.815		0.863		1.09	1.18			0.716
Yb	1.15	1.04	0.755		0.811	0.780		0.832		1.04	1.13			0.690
Lu	0.171		0.113		0.124	0.122					0.170			0.106
Pd	15.6	21.2	10.6	10.4	5.9	7.9	7.6	12.0	13.2	15.0	13.1	10.0	9.0	8.2
Ir	0.50	0.55	1.27	0.540	1.02	2.14	10.08	0.698	0.580	0.490	0.450	0.700	1.37	1.41
Pt	15.9		10.9			8.0		11.0	11.0	6.0	12.0			

Table 8.3 (continued). ZV-samples after Nisbet et al. (1987).

%	Z12	ZV1	ZV3	ZV4	ZV6	ZV7	ZV9	ZV10	ZV11	ZV12	ZV14	ZV71	ZV85	ZV87
SiO_2	47.39	49.68	50.17	49.23	49.68	47.70	47.31	47.41	47.51	47.07	51.03	48.36	48.07	49.86
TiO_2	0.33	0.42	0.44	0.40	0.40	0.32	0.30	0.30	0.31	0.30	0.48	0.36	0.35	0.44
Al_2O_3	6.61	8.61	9.02	8.14	8.24	6.70	6.29	6.36	6.36	6.17	9.64	7.57	7.16	8.94
FeO^*	10.89	11.59	11.65	11.37	11.63	10.87	10.73	10.69	10.87	10.54	11.75	11.13	11.24	11.61
MnO	0.18	0.20	0.20	0.19	0.19	0.18	0.18	0.18	0.18	0.18	0.19	0.19	0.18	0.19
MgO	26.65	19.11	17.67	20.81	20.20	25.90	27.63	27.18	26.96	28.29	15.98	23.14	24.41	18.41
CaO	6.81	8.87	9.17	8.43	8.27	7.00	6.59	6.83	6.78	6.36	9.40	7.81	7.37	9.12
Na_2O	1.00	1.42	1.58	1.34	1.27	1.25	1.17	0.98	0.98	1.05	1.42	1.34	1.16	1.34
K_2O	0.10	0.07	0.07	0.07	0.08	0.05	0.05	0.04	0.04	0.03	0.08	0.07	0.05	0.06
P_2O_5	0.02	0.03	0.03	0.03	0.03	0.02	0.02	0.02	0.02	0.02	0.03	0.02	0.02	0.03
(H_2O)	4.10	3.62	3.31	3.61	3.83	3.37	3.53	2.48	3.16	3.00	3.99	3.20	3.98	3.78

ppm	Z12
Rb	1.5
Sr	29
Ba	18
Zr	17
Sc	22
Cr	2600
Ni	1310
La	
Ce	
Nd	
Sm	
Eu	
Gd	
Dy	
Er	
Yb	
Lu	
Pd	11.9
Ir	1.70
Pt	

FeO^* total iron as FeO

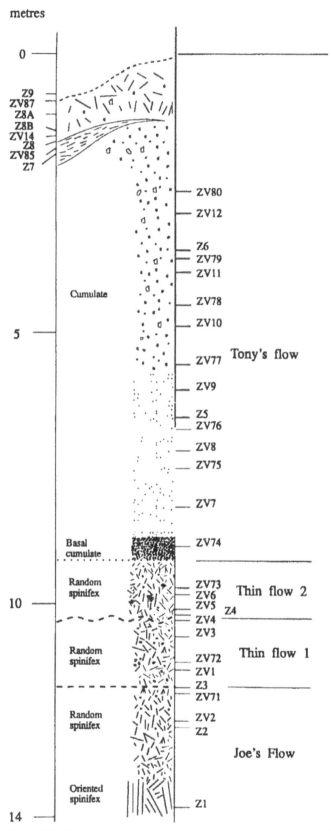

Figure 8.3. Diagrammatic section across surface outcrop of Joe's and Tony's flows showing sample locations of Z and ZV-series samples. After Renner (1989).

constraints on alteration available from these very fresh rocks to evaluate of the considerable uncertainties over the geochemical significance of the typically more altered komatiitic suites (e.g. Elthon 1986; de Wit & Hart 1986).

8.5 Z-SERIES SAMPLES: PETROGRAPHIC AND GEOCHEMICAL CONSTRAINTS ON ALTERATION OF KOMATIITES

The three flows exposed and sampled are illustrated in Figures 8.1, 8.2 and 8.3. The upper of the two spinifex flows (Tony's flow) is almost completely exposed and differentiated into a spinifex-textured upper part and a lower olivine-enriched part. Only the spinifex-textured unit of the underlying flow (Joe's flow) is exposed. The third sampled flow is a porphyritic unit stratigraphically above the spinifex flows. The drill-core recovered an almost complete section through the three flows as well as substantial sections above and below the flows (Fig. 8.2).

The freshness of the Z and ZV-series samples is immediately apparent in the proportion of olivine that has escaped serpentinization. In some sections (e.g. sample Z2, Appendix), equant solid olivine phenocrysts are mostly unaltered and even the finest skeletal blades (0.01 mm in width) contain remnants of the primary mineral (Fig. 8.4b). Pyroxene grains are similarly well preserved: The pigeonite centres of the larger skeletal pyroxene needles are unaltered, and in one example, an orthopyroxene core was identified (see below). Most remarkable, however, is the glass phase. Devitrified glass is common and some olivines have small silicate liquid inclusions of fresh unaltered glass (Nisbet et al. 1987). Microprobe analyses of some glass patches yield totals of up to 99.5% (Table 8.2). SEM investigations reveal that the glass contains abundant minute grains of an opaque mineral, but these grains are believed to be a final stage crystallization product and not a later devitrification or alteration phase.

The Z and ZV-series samples have all undergone some alteration as is reflected by their H_2O contents (2.3-4.4%, Table 8.3), by the presence of serpentine-filled and carbonate-filled veins spaced at ca. 10 cm to a metre, and by their oxygen, hydrogen and Rb-Sr isotopic systematics (Hegner et al. 1987). Carbonate-filled veins, only apparent in the unweathered drill core, are concentrated on flow boundaries.

Oxygen and hydrogen isotopic analyses of serpentines, mainly from rocks from the type section of the Reliance Formation and from locality B4, and thus more altered than the Z-series lavas, demonstrate that much of the serpentine formed (or last equilibrated) with meteoric water at low temperatures, although some serpentine fractions preserved isotopic compositions consistent with higher temperature greenschist facies equilibria (Kyser & Nisbet 1986; see Nisbet 1987). Whole-rock oxygen isotopic compositions from the Z-series Tony's flow range from $\delta^{18}O = 6.3$ to 7.5 in comparison to unaltered olivine compositions of $\delta^{18}O = 5.3$ (Nisbet et al. 1987). This range of whole-rock values is interpreted by Hegner et al. (1987) to imply low grade alteration of the groundmass. In contrast the oxygen isotopic compositions of olivines and clinopyroxenes appear unchanged.

8.5.1 Geochemical alteration

Even the limited alteration undergone by the Z-series lavas has changed the concentrations of some elements. We have used a variety of techniques to screen the analyses for post-emplacement alteration, including: 1) Comparison between the whole-rock chemistry and compositions constructed from modes and mineral compositions; 2) the repro-

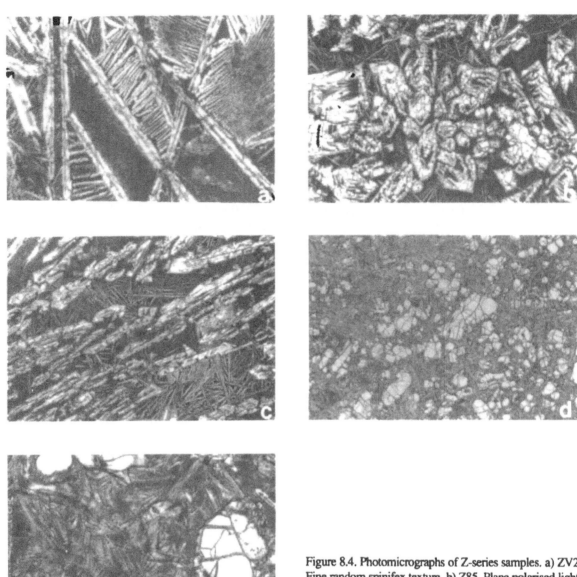

Figure 8.4. Photomicrographs of Z-series samples. a) ZV2. Fine random spinifex texture. b) Z85. Plane polarised light, B1 zone 7.5 mm across. c) ZV13. Plane polarised light, B1 zone 7.5 mm across. d) ZV78. Plane polarised light. Olivine cumulate. 7-0.5 mm across. e) ZV10. Plane polarised light. Detail of cumulate. 1.4 mm across.

186

duction of simple fractionation trends (e.g. olivine control) in layered flow units; 3) use of relict primary mineral compositions where appropriate partition coefficients are known or may be inferred from fractionation trends; and 4) comparison between less and more altered samples.

8.5.2 Petrography

The application of all the above techniques is based on petrographic interpretation of the compositions and crystallisation history of the lava flows. The upper portion of a typical layered spinifex-textured komatiite flow grades from a glassy flow top with sparse olivine phenocrysts (Fig. 8.4a) often with carbonate-filled veining, down into an increasingly coarse spinifex olivine zone. This comprises randomly orientated skeletal olivine (average Fo 91) occasionally passing into orientated, longer skeletal olivine plates, in a matrix containing skeletal pyroxene plates, small euhedral and skeletal chromite and an altered glassy matrix (Fig. 8.4b). The spinifex zone terminates against a layer termed the B1 zone, of tabular, skeletal olivine grains (Fo 91.3) oriented parallel to the plane of the flow (Fig. 8.4c, B_1 zone, c.f. Arndt et al. 1977). Below the B_1 zone the rock is rich in olivine phenocrysts (up to 50% of the rock) set in a fine-grained matrix of altered glass, skeletal clinopyroxene, equant chromite and a small proportion of skeletal olivine (Fig.

Table 8.4. Correlation with MgO Z-series samples.

Component	Error %[1] 2σ	MgO intercept[2] ± 2σ		Fo intercept[3]	MSWD[4]	No samples
Group 1						
MgO	2.5	–		–	–	–
TiO₂	2.0	49.0	1.5	90.2	2.6	28
Al₂O₃	2.0	49.4	1.0	90.7	0.8	28
FeO*[5]	2.0	–		91.2	2.1	28
SiO₂	1.0	–		88.2	0.9	28
Zr	5.0	46.4	6.0	86.6	13	15
Nd	5.0	47.9	2.6	88.7	1.0	7
Sm	5.0	49.4	2.9	90.7	0.6	7
Er	5.0	48.1	6.3	88.9	5.7	9
Sc	5.0	50.7	2.5	92.6	1.0	15
Group 2						
Na₂O	13	48.1	6.9		6.0	28
Na₂O[6]	13	51.4	7.0		2.9	25
Sr	5.0	45.2	3.3		4.1	15
Ce	5.0	44.6	6.2		8.9	6
Ce[7]	5.0	47.5	2.9		0.8	5
Mobile elements Group 3						
CaO	2.0	55.0	2.7		4.5	28
K₂O	5.0	38.1	5.4		86	28
Pd	10.0	39.6	6.3		18	15

1.Error estimated. 2.Intercept and error on MgO axis, regression after York (1969). 3.Forsterite content of olivine is that at intercept between regression line in MgO: Element plot, and regression line through olivine compositions. 4.Mean squared weighted deviate (York 1969). 5.Total iron as FeO. 6.Calculated excluding samples Z5, Z6 & Z8A. 7.Calculated excluding one sample.

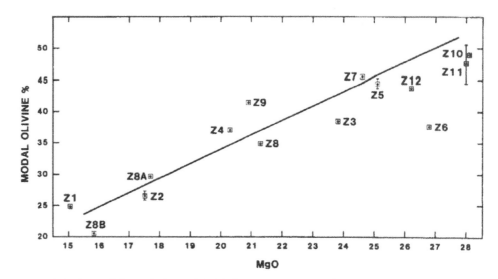

Figure 8.5. Modal olivine variation against MgO wt %, Z series samples.

Table 8.5. Modal compositions of Z-series lavas.

	Skeletal	Olivine total	Phenocryst	Cpx and glass	Chromite
Z1		24.7		75.3	0
Z2	26.0		0.6	73.1	0.4
Z3	21.8		16.7	61.1	0.4
Z4	26.1		10.9	62.1	0.9
Z5		44.3		55.1	0.6
Z6		37.6		60.9	1.5
Z7		45.6		53.4	1.0
Z8	32.6		2.1	65.0	0.4
Z8a	28.4		0.7	70.4	0.5
Z8b		20.1		79.5	0.4
Z9	24.6		18.2	56.6	0.5
Z10		48.9		50.5	0.6
Z11		47.5		51.6	0.6
Z12		43.7		55.6	0.6

8.4c, d). At the base of the flow the proportion of polyhedral olivine grain decreases and that of skeletal olivine and glass increases.

In the porphyritic B-zone (Pyke et al. 1973) olivine is predominantly equant and solid. These grains occur in clusters of a few to 10's of grains. Within the cumulate zone two types of equant grains can be distinguished on the basis of size and composition. Micro-phenocrysts (average Fo = 91.3, average diameter 0.3 mm) are abundant and comprise most of the cumulate grains. Larger equant grains (diameter > 0.6 mm, Fo = 93.0 to 93.6) comprise up to 5% of the cumulate grains. As will be shown below, the most fosterite-rich cores are too magnesian to have been in equilibrium with the erupted silicate liquid, indicating that these grains probably are xenocrysts.

The pyroxenes share the normal characteristics of pyroxenes in other Archaean komatiites and komatiitic basalts (Arndt & Fleet 1979). The smaller grains are dendritic and have Al-rich, Ca-poor augite compositions. Larger grains are complex, with margins of Ca-poor augite, central portions of Mg-rich pigeonite, and in some cases, cores of bronzite (Table 8.2). Although bronzite has not previously been reported in spinifex rocks, it is not uncommon in rapidly crystallized pyroxenes in lunar basalts. In most komatiites, orthopyroxene cores to complex pyroxene needles probably have altered to secondary minerals.

The isotropic glass samples are of evolved liquids with the composition of a high-Al basalt with 4-6% MgO and 15-18% Al_2O_3. These compositions are similar to those in the Mesozoic Gorgona komatiites (Echeverria 1982), except that the Belingwe rocks have lower CaO (9% in the Belingwe rocks, compared with 15% in the Gorgona samples). The chrome-spinel grains are zoned slightly towards Fe-rich margins.

From the petrography – the minerals identified and their crystallographic habits – it appears that the only liquidus phases present during the bulk of the cooling interval of the Z-series flows were olivine and minor chromite. Only these phases can have been involved in any mineral fractionation or redistribution that took place during eruption. In the modal data presented in Table 8.5 separate values are given for equant and skeletal olivine. This was done so that those rocks with a significant component of phenocryst or cumulate olivine could be distinguished from those with only skeletal olivine. Only the samples without phenocryst olivine can be considered as possible liquid compositions. Of these, the most magnesian is sample Z8 with 21.6% MgO but adjacent specimens Z28A and Z28B are less magnesian with 18 and 16% MgO respectively.

8.5.3 *Comparison of modes*

Figure 8.5 plots total modal olivine against MgO content. The positive correlation supports the idea that the MgO contents are a primary magmatic feature of the lavas but the uncertainty in the glass composition and differing degrees of crystallinity means that this correlation cannot be used to provide a precise control on alteration of MgO concentrations.

8.5.4 *Fractionation trends*

The coherence of geochemical trends with concentrations recalculated on an anhydrous basis provides the main evidence that extensive changes have not taken place in the concentrations of many elements in komatiite suites (e.g. Nesbitt & Sun 1976; Arndt et al. 1977; Nisbet et al. 1977). However, a coherent geochemical trend may also be a consequence of a systematic alteration of an originally coherent pattern (Smith & Erlank 1982). In the Z and ZV-series lavas the compositions of the controlling phenocryst phases are known, and thus it is considered unlikely that such trends would remain unshifted on two element plots after significant alteration. This provides an additional constraint on the extent of element migration during alteration.

Figure 8.6 portrays the Z and ZV-series chemical data plotted against MgO on variation diagrams. The MgO contents of phenocryst-poor chill and spinifex lavas are in the range 15-22% MgO, and those of porphyritic and cumulate rocks from 20-28% (Table 8.3; Appendix). The variation diagrams reflect both cumulation and fractionation processes. Samples of the chilled margins (e.g. Z3) average 25.8% MgO which is identical to the depth-averaged composition of Joe's and Tony's flows of 25.7% MgO (Renner 1989). However the chilled margin samples contain a significant proportion of the equant Fo ≈ 93 olivine phenocrysts (e.g. Z3 with 16.9 vol.%, Table 8.5). Subtraction of these implies that the silicate liquid that formed the sampled portion of the flow had 18-20%

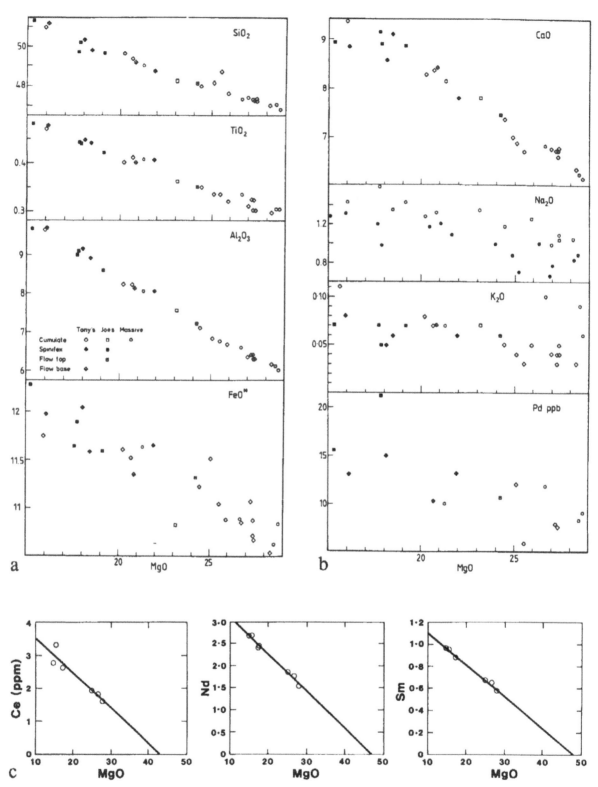

Figure 8.6. Variation diagrams against MgO, Z and ZV-series samples. Analyses recalculated anhydrous. a) SiO$_2$, TiO$_2$, Al$_2$O$_3$ + FeO. b) CaO, Na$_2$O, K$_2$O + Pb, symbols as in a. c) Ce, Nd, Sm.

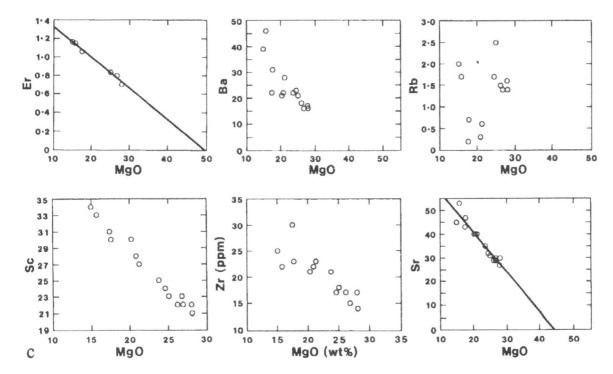

Figure 8.6. (continued). c) Er, Ba, Rb, Sr, Sc, Zr.

MgO on eruption. Renner (1989) interprets the petrogenesis to involve derivation of a parental komatiitic liquid with ca. 25.7% MgO which crystallised 17 to 19 vol.% olivine phenocrysts before emplacement. The phenocryst-bearing magmas ponded at a stage when the liquid containing about 20% MgO and fractionated in situ.

Samples of all three flows fall on the same chemical variation trends, except in the case of LIL elements which appear to be enriched in the uppermost porphyritic flow (Z10, Z11, Z12). In many of the diagrams, the data fall on closely constrained trends that project directly towards plausible olivine compositions. Table 8.4 lists the regression lines, mean squared weighted deviates (MSWD) and the olivine intercept composition of these trends. Regressions are calculated using the York (1969) routine which minimises least-squares deviates weighted for errors in both x and y parameters as given in Table 8.4. For incompatible elements, the MgO content of the fractionating minerals (olivine and minor chromite) is given by the MgO axis intercept. For compatible elements the MgO content is that at the intercept of rock and mineral composition trends (e.g. Fig. 8.7).

The elements divide into three groups. The first group (Al, Ti, Sc, Nd, Sm, Er, Si, Fe Ni) are highly correlated with MgO and give tightly grouped olivine intercept compositions (mean of Ti and Al intercepts = Fo 90.5 ± 1.2, 2σ). This is consistent with the spinifex phenocryst compositions. The trends are interpreted as preserved unaltered concentration data. A second group (Na_2O, Sr and Ce) is less well correlated with MgO but has olivine intercept compositions within error of the mean of the first group and this group may still reflect olivine control. The scatter may either reflect the lower precision of the analyses, limited alteration or both. A third group (Ca, K, Rb, Ba) either exhibit complete scatter (K, Rb, Ba) or trends which deviate from the expected olivine control (Ca, Ba) and are interpreted to have undergone more substantial alteration. This alteration was systematic for CaO which is highly correlated with MgO.

191

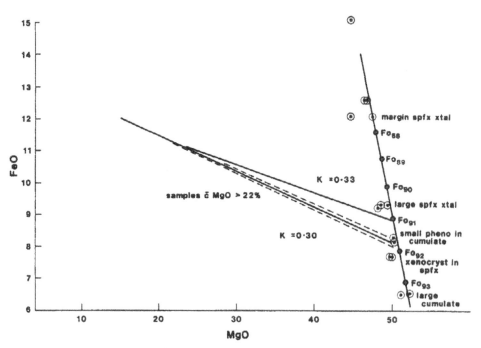

Figure 8.7.a) Calculated fractionation trends Z and ZV-series samples. FeO:MgO variation for Z and ZV series lavas (Table 8.4). Solid line with dashed lines is the regression line and 2σ confidence limits. Intercept is at the olivine composition Fo 91.4 (Table 8.2). If this trend is controlled by olivine fractionation/cumulation in equilibrium with 20% MgO lava, this implies $K_D^{Fe:Mg} = 0.29$. Tie lines to olivine Fo 90.3 appropriate for equilibrium with 20% MgO lava if $K_D^{Fe:Mg} = 0.33$ are also shown.

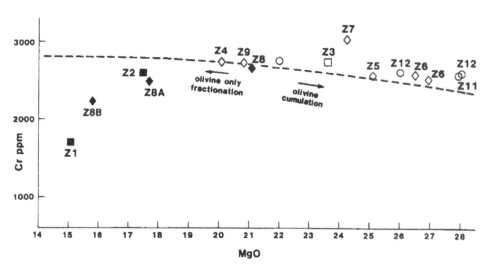

Figure 8.7.b) Calculated fractionation trends Z and ZV-series samples. Cr:MgO variation. Dashed line is trend for olivine factionation/cumulation from 22% MgO liquid with $D_{Cr} = \dfrac{16.4}{MgO} - 0.06$.

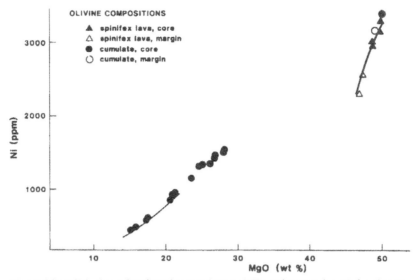

Figure 8.7.c) Calculated fractionation trends Z and ZV-series samples. Ni fractionation and best-fit theoretical curves though lava and olivine compositions calculated for $D_{Ni} = \dfrac{125}{MgO} - 2.25$ (see Arndt 1986).

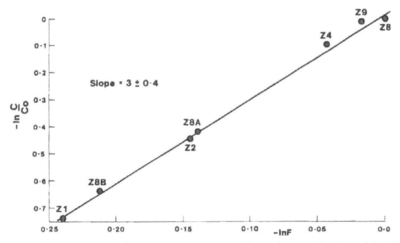

Figure 8.7.d) Calculated fractionation trends Z and ZV-series samples. Plot of $-\ln(C/C_0)$ versus $-\ln F$ where C/C_0 is ratio of Ni concentration to that in 22% MgO parent liquid and F is proportion of fractionated liquid calculated from MgO content.

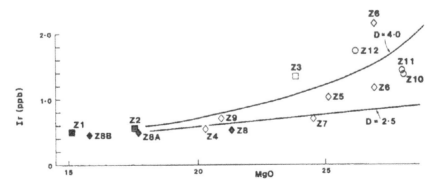

Figure 8.7.e) Calculated fractionation trends Z and ZV-series samples. Ir:MgO variation and fractionation trends calculated for D_{Ir} (olivine: liquid) equal to 2.5 and 4.0.

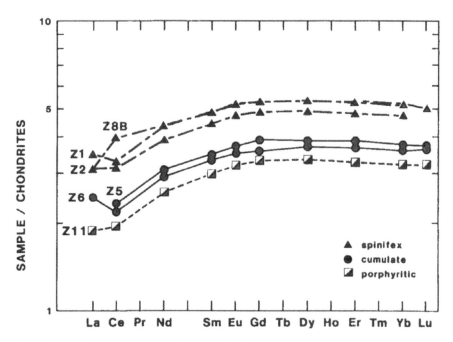

Figure 8.8. REE patterns normalised to chondrite, Z-series samples.

Most of the REE appear immobile, forming tight trends and projecting towards 'reasonable' olivine compositions. Chondrite-normalized REE patterns are largely smooth and show LREE depletion, like many other komatiites of this age (Fig. 8.8). The exception is Ce, which has an anomalous MgO intercept (44.7%), and shows pronounced anomalies in the chondrite-normalized plots: In most samples the anomaly is negative, but in one (sample Z8b) it is positive. These anomalies cannot easily be explained by igneous processes and likely result from alteration. Ce^{3+} can be oxidized to Ce^{4+} and under these conditions it is mobile. Ce anomalies are not seen in komatiites from other areas; this suggests, perhaps, that the alteration that affected these otherwise well-preserved lavas was associated with recent fluid movement.

The abundances of Na and K (but not Rb or Sr) in samples Z10-12 are systematically higher than in samples from the lower two flows. It is not clear whether this is due to mobility of these elements, or to a difference in the initial magma compositions. No systematic difference is seen in the concentrations of immobile incompatible elements like Nd or Sm.

For the minor elements compatible with olivine and chromite, it is not so simple to decide whether or not an element was mobile, because calculation of the theoretical fractionation trends requires an estimation of partition coefficients that in many cases are known only very approximately. However, Ni, Cr and Ir fall on relatively tight trends and vary in a manner consistent with partition coefficients within acceptable ranges. They too appear to have been immobile.

The behaviour of elements during the alteration of these flows is similar to that deduced from studies of more altered Archaean komatiites and basalts (Smith & Erlank 1982; Beswick 1982; Arndt & Nesbitt 1982). In general, those elements identified as mobile in this study also show marked deviations from coherent behaviour in the studies of the more altered lavas. In this context, it is important to note that some of the elements shown to be mobile on the basis of inappropriate MgO axis intercepts have moderately

194

tightly constrained trends (e.g. CaO for which MSWD = 4.5). The case of CaO is particularly interesting because for these relatively fresh Belingwe lavas the trend projects to a high MgO intercept (55%) whereas for the more altered komatiitic lavas from other areas in Belingwe (Nisbet et al. 1977), in Barberton (Smith & Erlank 1982) and Munro Township (Arndt & Nesbitt 1982), the intercepts are low (40-47%). The explanation for this difference is not known.

8.5.5 *Composition of fractionating phases – Partition coefficients*

The fractionation trends can be inverted to give olivine: Liquid distribution coefficients (except for Cr which largely partitions into chromite). These may be compared with distribution coefficients derived from experimental work and the compositions of the olivine phenocryst phases.

Figure 8.7a illustrates the FeO:MgO correlation and its intercept with the olivine composition trend at Fo 91.2±0.5 (2σ) which is within error of the mean of the well-defined TiO_2:MgO and Al_2O_3:MgO intercepts of Fo 90.5±1.2 (MgO = 49.2± 0.9). A 20% MgO liquid (FeO*=11.6% from the regression) in equilibrium with Fo 91.2 olivine implies $K_D^{Fe:Mg} = 0.29$. This is within the range of experimentally determined K_D values given the uncertainty of the Fe_2O_3 contents of the liquids (Bickle 1982). If $K_D^{Fe:Mg} = 0.33$ the predicted equilibrium olivines would be Fo 90.3 in a 20% MgO liquid, but control by such olivine compositions lies outside the error bounds on the regression (Fig. 8.7a). Accumulation of the more magnesian Fo 93 olivine xenocrysts (maximum of 18% in Z9) would not be detectable within the errors of the FeO:MgO correlation.

Partition coefficients for minor or trace elements can be derived from the slope of log (C_l/C_0) versus log F where C_0 is the concentration in the hypothetical initial melt, C_l the concentration in the fractionated liquid and F the proportion of fractionated liquid (Cawthorn & McIver 1977). Such plots for Ni and Ir are illustrated in Figures 8.7c, d, e compared with partition coefficients calculated from olivine compositions in Table 8.6. The estimated partition coefficient for Ni of 3±0.4 for a 22% MgO liquid agrees well with the experimental data of Arndt (1977) and data reviewed by Bickle (1982).

8.5.6 *Comparison with the more altered rocks*

The consistency of the geochemical evidence for olivine control, as shown by plots of SiO_2, TiO_2, Al_2O_3, NaO, FeO, Na_2O, NiO, Sr, Zr, Y, Sc, Ir and most rare earth elements against MgO, and the consistency of Fe:Mg and Ni distribution coefficients with the compositions of phenocryst phases and experimentally determined values, all argue that

Table 8.6. Partition coefficients. Calculated from whole rock and olivine compositions and from fractionation trends.

Element	From olivine compositions for 22% MgO liquid		From fractionation trend after Cawthorn & McIver (1977) (for 22% MgO liquid)	
	D	± 2σ	D	± 2σ
Ni	3.4	0.4	3	0.8
Ir	3.3	1.4		
Ca	0.03	0.01		

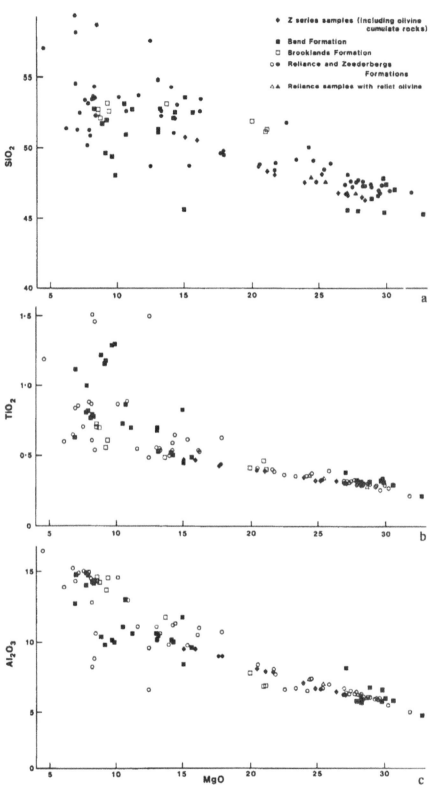

Figure 8.9. Geochemical variation against MgO, Belingwe lavas. Analyses from Bickle et al. (1975), Nisbet et al. (1977) and Tables 8.3 and 8.7. Apart from Z-series samples, rocks obviously enriched in cumulate olivine excluded. Analyses recalculated anhydrous. Zr:MgO plot shows olivine and olivine-clinopyroxene fractionation trends with olivine-clinopyroxene ratio estimated from CaO:MgO variation. a) SiO_2. b) TiO_2. c) Al_2O_3.

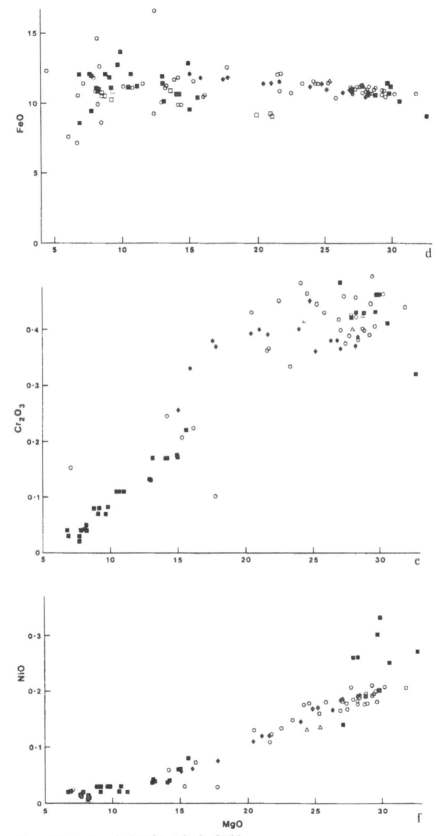

Figure 8.9. (continued). d) FeO*. e) Cr₂O₃. f)NiO.

197

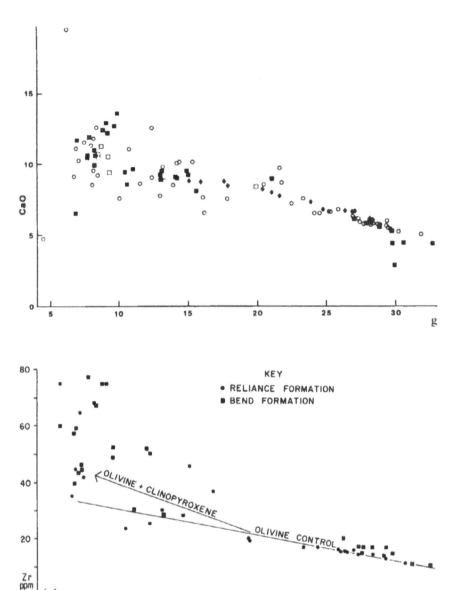

Figure 8.9. (continued). g) CaO. h) Zr.

significant alteration has not occurred in the concentrations of these elements in the Z-series flows. This permits comparison between the chemistry of the Z-series lavas and the chemistry of the more altered, completely serpentinised, lavas from elsewhere within the Reliance Formation (e.g. Nisbet et al. 1977) to assess the consequences of chemical alteration on komatiite compositions.

Figure 8.9 compares chemical variation of the Z-series suites with the other Reliance Formation samples from Belingwe. TiO_2, Al_2O_3, FeO^*, Cr_2O_3, and NiO exhibit no significant differences. SiO_2 and CaO in the 20-25% MgO Z-series lavas are slightly lower than in completely serpentinised Reliance lavas of equivalent MgO, but not outside the scatter in SiO_2 and CaO contents of the more altered lavas. Rare earth element concentrations and patterns in the more altered lavas (Hawkesworth & O'Nions 1977) also are similar to the Z-series compositions (Fig. 8.8). The only marked difference is for Na_2O. The Z-series lavas exhibit Na_2O:MgO variation within error of olivine control albeit with some scatter. The more altered lavas from elsewhere in the Reliance Formation exhibit much lower Na_2O contents (< 0.3% versus 0.8 to 1.2% in the Z-series lavas) and no significant correlation with MgO (Nisbet et al. 1977). This is illustrated in Figure 8.10 in which the Z-series lavas are compared with a global suite of generally more altered samples. The majority of the more altered lavas thus appear to have lost substantial Na_2O.

8.6 MgO CONTENTS OF PARENT LIQUIDS

The Z-series komatiites and the completely serpentinised Reliance komatiites have concentrations of SiO_2, TiO_2, Al_2O_3, FeO, NiO, Cr_2O_3, Sc, and Zr and REE (except Ce) and platinum group elements which are little perturbed by alteration. Sr and Na_2O concentrations are little changed in the Z-series lavas but substantially reduced in the completely serpentinised rocks. CaO displays small (but opposite) systematic shifts in both suites. K, Rb and Ba are substantially perturbed in all the rocks. These observations have the following implications.

The preservation of olivine control lines in MgO-oxide variation diagrams indicates that the high MgO concentrations of komatiites are primary and not the result of metasomatic enrichment as suggested by de Wit et al. (1986). The spinifex-textured Z-series samples crystallised from liquids with 18-22% MgO (as indicated by the compositions of the most Mg-rich spinifex textured samples Z2, Z8A, 28, and the compositions of the most magnesian spinifex olivines within these samples which are in equilibrium with liquids containing 18-20% MgO). The Z-series spinifex samples are petrographically similar to the least magnesian completely serpentinised Reliance komatiite samples. On this basis we interpret the most magnesian Reliance spinifex samples B4 and the numerous skeletal hopper olivine samples with 25-32% MgO as representing original lava compositions.

The constraint on the MgO contents of the liquids is less good given the scarcity of spinifex-textured samples and samples with relict olivine. Spinifex samples B4 with Fo 92.6 olivine would have crystallised from a liquid with between 23.5 to 25.5% MgO depending on the choice of $K_D^{Fe:Mg}$ (i.e. 0.30 to 0.33, Bickle 1982; Arndt 1986). It is of interest that the Z-series lavas contain xenocrysts of Fo 92 to 93.5 suggesting that the magma fractionated from a liquid as magnesian as 25 to 28.5% MgO, depending on the appropriate $K_D^{Fe:Mg}$, and then cooled quickly enough for the xenocrysts not to re-equilibrate with the magma.

8.7 REGIONAL SUITES FROM BELINGWE – CHEMISTRY AND COMPARISONS

8.7.1 *Komatiite suites*

Three distinct komatiite suites each consisting of komatiites and basalts have been sampled: 1) The Reliance Formation suites (discussed above); 2) the Bend Formation suite from the Mtshingwe Group; and 3) a few samples from the Brooklands Formation of the Mtshingwe Group.

Major and trace element concentrations, recalculated on an anhydrous basis, are presented in Tables 8.7 and 8.8. Figure 8.9 illustrates the chemical variation, with major basalts (MgO < 15%) and komatiites (MgO > 20%) is clearly visible. The komatiites exhibit very coherent variation trends with MgO and their compositions and petrogenesis are discussed below.

The major and trace element trends displayed by the Reliance Formation komatiites have already been discussed in detail (Bickle et al. 1976, 1977; Nisbet et al. 1977) and the new data are consistent with that previously published. The Bend Formation komatiites exhibit a slightly more restricted range of MgO contents but whether this represents sampling bias or is representative of erupted compositions is uncertain: The Bend Formation samples differ from the Reliance Formation samples in being collected entirely from layered spinifex textured flows (Fig. 4.6), such flows being relatively rare in the Reliance Formation except at the Z-series locality (Fig. 8.1). The Bend Formation compositions are very similar for most elements to the Reliance Formation compositions although variation trends are slightly less coherent. The scatter is possibly a function of variation in degree of alteration between the two suites. It is not possible quantitatively to compare the relative alteration states of the two suites but it is notable that the few samples with relict olivine occur entirely within the Reliance Formation.

Table 8.7. Major and trace element analyses of Mtshingwe Group lavas.

| | Hokonui Formation | | | | | | |
	GC11	RL7549	RL7544	RL7545	RL7541	GC18	SL582
SiO_2	55.49	60.38	66.25	66.30	69.32	71.93	79.09
TiO_2	0.913	0.686	0.62	0.64	0.60	0.25	0.14
Al_2O_3	15.28	17.56	15.79	16.19	14.99	13.72	12.19
FeO*	9.37	5.88	4.78	4.75	4.21	2.75	0.81
MnO	0.13	0.09	0.07	0.07	0.06	0.03	0.02
MgO	4.70	3.30	2.79	2.74	2.29	1.17	0.26
CaO	6.63	3.03	2.01	2.02	2.36	2.92	0.14
Na_2O	2.38	5.35	5.08	5.02	5.90	2.07	4.99
K_2O	1.78	1.71	1.39	1.62	0.53	4.48	2.70
P_2O_5	0.29	0.14	0.13	0.13	0.12	0.08	0.04
Cr_2O_3	0.02	0.01	0.01	0.01	0.01	0.01	0.00
NiO	0.01	0.01	0.004	0.00	0.01	0.00	0.00
Total	97.83	98.70	99.36	99.91	100.80	99.66	100.50
LOI	2.11	4.53	3.49	3.50	3.22	1.15	2.19
FeO**	9.03	4.04	3.96	3.53	3.40	1.95	0.96
Ba	596	641	511	584	309	828	409
Y	23.9	20	16.2	17	8	21.1	15
Zr	183	164	163	167	38	165	166
Nb	8.7	7.1	5.7	5.6	0.8	6.5	5.4
Rb	53	72	60	69	4	121	55
Sr	305	207	162	157	143	190	63
Ti/Zr	30	25	23	24	97	9.1	5
Ti/Y	229	179	229	193	409	71	49
Trace analyses	L	ETH	L	ETH	ETH	L	ETH

Table 8.7 (continued).

	Bend Formation										
	Cumulates		Komatiites								
	SL708	SL413	SL585	SL445	SL501	SL442	SL504	SL741	SL502	SL703	SL550
SiO$_2$	46.28	43.96	45.33	47.12	47.41	45.46	47.86	46.46	47.49	45.52	45.57
TiO$_2$	0.10	0.19	0.22	0.30	0.32	0.34	0.33	0.32	0.32	0.33	0.39
Al$_2$O$_3$	2.54	4.02	4.88	5.92	6.07	6.65	5.86	6.85	5.76	5.83	8.21
FeO*	8.51	10.52	9.08	10.11	11.15	10.66	11.37	10.54	10.51	11.28	10.94
MnO	0.06	0.17	0.15	0.18	0.19	0.16	0.16	0.16	0.19	0.20	0.17
MgO	41.57	35.43	32.65	30.54	29.93	29.76	29.67	28.83	28.15	27.85	27.02
CaO	0.12	2.46	4.39	4.48	2.98	4.39	5.24	5.53	5.87	5.78	6.12
Na$_2$O	0.00	0.05	0.04	0.00	0.00	0.00	0.00	0.00	0.28	0.21	0.08
K$_2$O	0.00	0.00	0.18	0.00	0.00	0.00	0.00	0.20	0.00	0.00	0.15
P$_2$O$_3$	0.03	0.04	0.04	0.04	0.04	0.04	0.04	0.04	0.04	0.04	0.05
Cr$_2$O$_3$	0.24	0.80	0.32	0.41	0.46	0.46	0.43	0.43	0.43	0.42	0.48
NiO	0.48	0.25	0.27	0.25	0.33	0.20	0.30	0.19	0.26	0.26	0.14
Total	100.50	98.77	98.20	100.19	99.78	99.00	102.20	100.50	100.20	98.85	100.30
LOI	11.70	11.16	8.60	8.93	8.68	8.82	7.59	8.05	7.03	7.76	7.57
FeO**	3.72	5.58	4.48	4.48	4.14	4.86	4.35	6.72	4.77	5.59	7.28
Ba	–	30	61	105	71	90	32	32	23	72	57
Y	2.6	3.9	4.8	6.4	6.2	6.6	6.2	6.7	5.1	6.4	7.6
Zr	5.8	9.0	11.6	16.3	15.0	18.6	18.0	17.3	18.4	15.8	21.6
Nb	0.0	0.2	0.2	0.6	0.4	0.6	0.3	0.4	0.4	0.8	0.7
Rb	0.9	2.0	16.2	3.6	7.8	4.4	0.8	23.1	1.2	3.8	19.5
Sr	0.3	8.7	11.2	7.7	8.5	7.0	2.6	9.0	4.3	6.7	8.4
Ti/Zr	103	127	114	110	128	110	109	110	106	128	114
Ti/Y	231	292	275	279	309	311	310	285	112	315	275
Trace analyses	L	L	L	L	L	L	L	L	L	L	L

	Bend Formation										
	Formation lavas										
	Komatiitic basalts and basalts										
	SL752	SL776	SL745	SL780	SL782	SL781	SL720	SL719	SL785	SL740	SL702
SiO$_2$	52.55	53.61	45.54	52.55	52.18	52.86	51.12	51.35	52.78	53.15	50.90
TiO$_2$	0.49	0.45	0.83	0.51	0.52	0.53	0.68	0.70	0.70	0.73	0.87
Al$_2$O$_3$	9.64	8.46	11.81	10.03	10.15	10.49	10.17	10.59	10.63	11.10	13.04
FeO*	10.40	9.56	12.81	10.66	10.66	10.16	11.43	11.90	11.21	11.19	12.01
MnO	0.16	0.16	0.21	0.17	0.17	0.15	0.18	0.19	0.18	0.17	0.20
MgO	15.55	14.98	14.90	14.20	14.13	13.12	12.98	12.98	11.07	10.44	10.59
CaO	8.06	9.21	9.55	8.97	9.02	9.49	8.87	9.25	9.55	9.40	8.66
Na$_2$O	2.22	2.26	1.07	2.28	2.43	2.47	2.29	2.15	2.63	2.80	2.94
K$_2$O	0.00	0.00	0.31	0.00	0.00	0.00	0.11	0.11	0.26	0.17	0.11
P$_2$O$_5$	0.06	0.05	0.08	0.06	0.06	0.06	0.08	0.08	0.08	0.08	0.09
Cr$_2$O$_3$	0.22	0.17	0.18	0.17	0.17	0.17	0.13	0.13	0.11	0.11	0.11
NiO	0.08	0.06	0.06	0.04	0.04	0.04	0.04	0.04	0.02	0.02	0.03
Total	100.40	99.82	98.52	99.73	99.61	99.63	99.11	99.55	100	100.30	100.60
LOI	3.59	2.80	3.98	1.18	4.03	3.69	2.11	3.99	2.30	0.63	2.76
FeO**	8.10	7.43	10.35	7.76	7.61	7.13	8.13	8.54	8.04	7.10	9.27
Ba	48	40	226	305	51	108	109	116	142	112	101
Y	10.6	–	16	12	11	11.2	13.9	14.8	17	15.4	17.4

201

Table 8.7 (continued).

	Bend Formation Formation lavas Komatiitic basalts and basalts										
	SL752	SL776	SL745	SL780	SL782	SL781	SL720	SL719	SL785	SL740	SL702
Zr	29.8	–	52	34	32	32	52	54	57	53	51
Nb	0.6	–	0.5	1.5	0.5	1.1	2.3	2.2	1.0	2.0	1.7
Rb	1.6	–	2.1	1.2	0.0	0.5	1.5	1.4	3.6	1.9	2.4
Sr	57	–	35	121	138	156	28.4	36	106	75	78
Ti/Zr	99	–	96	93	101	99	78	78	76	83	102
Ti/Y	277	–	311	226	253	284	293	284	215	284	300
Trace analyses	L	L	ETH	ETH	ETH	L	L	L	ETH	L	L

	Bend Formation Komatiitic basalts and basalts									
	SL716	SL714	SL713	SL715	SL526	SL734	SL736	SL737	SL403	SL498
SiO_2	48.01	49.46	49.67	52.04	51.75	53.58	52.82	53.56	51.30	50.15
TiO_2	1.30	1.29	1.16	1.18	1.22	0.78	0.79	0.79	0.82	1.00
Al_2O_3	10.01	10.14	9.79	9.80	10.39	14.20	14.23	14.25	14.83	14.93
FeO*	13.67	12.73	11.81	11.11	12.08	10.92	10.80	10.98	11.92	11.98
MnO	0.23	0.22	0.21	0.20	0.20	0.17	0.16	0.17	0.18	0.19
MgO	9.83	9.65	9.08	9.14	8.78	8.26	8.21	8.17	7.78	7.73
CaO	13.65	12.76	12.90	12.20	12.47	10.59	9.89	10.98	11.93	10.58
Na_2O	0.92	0.96	1.55	1.77	1.99	1.70	2.18	1.46	1.16	2.49
K_2O	0.44	1.26	1.17	1.41	0.17	0.12	0.04	0.10	0.09	0.05
P_2O_3	0.11	0.11	0.10	0.10	0.11	0.07	0.07	0.07	0.08	0.10
Cr_2O_3	0.08	0.07	0.07	0.08	0.08	0.04	0.05	0.04	0.04	0.03
NiO	0.03	0.03	0.02	0.03	0.03	0.01	0.01	0.01	0.02	0.01
Total	99.51	99.83	97.64	100	100	101.40	100.20	101.60	101.20	100.30
H_2O	1.86	1.67	1.73	1.07	1.58	3.57	3.55	2.62	2.29	2.40
FeO**	9.89	9.34	8.40	8.39	8.36	7.70	8.06	7.49	8.89	7.54
Ba	203	81	461	556	129	92	71	81	70	83
Y	19.2	18.4	17.5	16.6	18.0	14.7	15.4	15.3	15.7	19.0
Zr	77	77	69	70	79	46	48	46	41	60
Nb	4.4	4.5	4.5	3.9	4.6	1.1	2.1	2.1	1.4	2.4
Rb	4.5	10.8	9.4	12.3	1.8	4.8	1.7	3.8	3.4	1.9
Sr	232	221	215	146	280	69	75	75	7.2	50
Ti/Zr	101	100	101	101	93	102	99	103	120	100
Ti/Y	406	420	397	426	406	318	308	310	313	316
Trace analyses	L	L	L	L	L	L	L	L	L	L

				Koodoovale Formation Felsic agglomerates				
	SL499	SL739	SL461	SL733	SL729	SL730	SL728	SL731
SiO_2	53.22	58.24	51.32	63.00	64.39	66.98	72.78	75.66
TiO2	0.81	0.63	1.12	0.64	0.49	0.62	0.48	0.18
Al_2O_3	14.07	12.75	14.77	17.49	15.50	15.98	14.29	14.42
FeO*	9.43	8.58	12.05	3.73	6.34	2.66	1.56	1.50

Table 8.7 (continued).

				Koodoovale Formation Felsic agglomerates				
	SL499	SL739	SL461	SL733	SL729	SL730	SL728	SL731
MnO	0.14	0.15	0.18	0.09	0.29	0.07	0.03	0.03
MgO	7.69	6.83	6.87	3.52	2.59	2.60	1.02	0.43
CaO	10.53	6.50	11.69	4.46	5.36	3.72	3.62	1.77
Na$_2$O	1.92	5.49	1.58	4.66	2.89	4.90	5.37	5.16
K$_2$O	0.20	0.29	0.07	1.47	1.07	1.16	0.49	1.38
P$_2$O$_5$	0.10	0.09	0.11	0.31	0.17	0.28	0.24	0.07
Cr$_2$O$_3$	0.02	0.04	0.03	0.01	0.00	0.01	0.01	0.00
NiO	0.01	0.02	0.02	0.01	0.00	0.01	0.00	0.00
Total	99	100.40	100.96	99.74	99.72	99.25	99.90	100.60
H$_2$O	1.65	1.58	2.31	4.67	3.79	3.51	2	1.48
FeO**	6.01	6.31	7.91	2.71	3.99	1.85	1.07	0.95
Ba	113	223	94	885	495	737	231	464
Y	18.1	14.9	24	14.1	11.2	12.7	10.3	3.8
Zr	60	61	75	166	121	141	124	103
Nb	2.0	2.1	1.1	4.2	3.0	3.6	2.1	1.8
Rb	8.7	3.8	0.3	49	32	38	14	38
Sr	138	64	107	595	232	544	551	291
Ti/Zr	81	62	92	23	24	26	23	10
Ti/Y	268	253	249	272	262	293	279	284
Trace analyses	L	L	ETH	L	L	L	L	L

	Brooklands Formation Komatiites			Komatiitic basalts					
	NG728	NG731	NG730	NG780	NG771	NG772	NG775	NG776	NG777
SiO$_2$	51.37	51.23	51.97	53.18	52.74	53.25	52.16	52.77	52.46
TiO$_2$	0.40	0.47	0.42	0.49	0.61	0.56	0.70	0.71	0.72
Al$_2$O$_3$	6.92	6.91	7.85	11.80	14.60	13.73	14.23	14.35	14.58
FeO*	9.04	9.21	9.13	10.85	10.80	10.24	10.51	10.52	10.70
MnO	0.18	0.18	0.18	0.18	0.17	0.18	0.18	0.17	0.19
MgO	21.01	20.93	19.87	13.55	9.28	9.22	8.66	8.49	8.52
CaO	8.96	8.91	8.39	7.60	9.41	10.53	11.25	10.70	10.49
Na$_2$O	1.02	1.02	1.03	1.02	1.02	1.01	1.00	0.99	1.01
K$_2$O	0.02	0.03	0.05	0.02	0.06	0.05	0.03	0.04	0.05
P$_2$O$_3$	0.04	0.04	0.05	0.04	0.05	0.05	0.06	0.05	0.05
Total	98.52	97.84	97.21	97.82	98.29	99.52	99.82	100.69	99.22
H$_2$O	4.48	8.27	5.05	3.48	7.63	6.02	5.93	1.08	7.66

*Total iron as FeO. **Analyzed FeO. Analyses recalculated on anhydrous basis. Total: Total analyses on (dry) XRF fused disc. L = trace element analysis (Y, Nb, Zr, Rb, Sr) at Leeds, ETH = trace element analysis at ETH Zurich. Most major element data done at ETH Zurich by XRF, some by wet chemistry at Cambridge (J. Scoon). Brooklands analyses done at Oxford by XRF. Total is total on oxidised fused disc. Results presented here with Fe as FeO. Where analysed, FeO is also given.

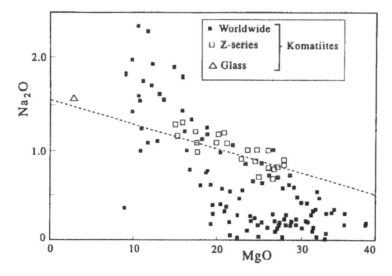

Figure 8.10. Variation diagram in which Na₂O contents of the relatively fresh Z-series komatiites from the SASKMAR location are compared with those of a global suite of generally more altered samples. The dashed line passes through the composition of olivine that fractionated in the flow, and close to the composition of glass in an olivine inclusion (Nisbet et al. 1987).

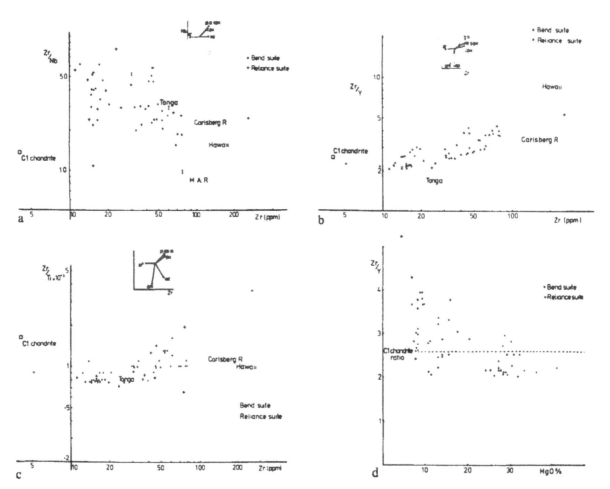

Figure 8.11. Trace element plots showing chondrite ratios, Belingwe komatiite-basalt sequences. Chondrite values, and insets showing fraction trends, after Pearce & Norry (1979). a) Zr/Nb:Zr. b) Zr/Y:Zr. c) Zr/Ti:Zr. d) Zr/Y:MgO.

Table 8.8. MgO (wt%), Nb, Zr, Y, Sr, Rb, Ti (ppm) analyses recalculated anhydrous.

Reliance komatiites Sample no.	MgO	Nb	Zr	Y	Sr	Rb	Ti	Ti/Zr	Ti/Y
B4	27.9	0.3	15.7	7.2	29.8	2.3	1807	115	251
NG152	31.8	0.2	11.0	5.4	9.2	1.6	1324	120	245
NG207	27.1	0.3	15.1	7.0	16.3	17.0	1918	127	274
NG7618	30.2	0.0	12.8	5.6	24.8	11.2	1648	129	294
NG7619	29.4	0.3	14.0	6.8	21.0	11.0	1781	127	262
NG7620	27.3	0.7	15.3	6.9	14.7	17.2	1900	124	275
NG7621	20.5	0.5	19.2	6.7	26.6	18.6	2500	130	373
NG7622	27.4	1.4	15.2	7.0	12.7	20.3	1948	128	278
NG7617	28.3	0.6	14.1	6.7	19.5	11.1	1822	129	272
NG7631	26.9	0.4	15.6	7.6	16.1	21.1	1960	126	255
NG7638	24.4	0.6	16.8	7.7	32.0	1.8	2182	130	283
NG7639	25.4	0.4	15.9	7.3	28.7	1.4	2056	129	282
Mean								126	279
Reliance basalts and komatiitic basalts									
NG120		1.7	36.6	12.0	11.7	0.6	3777	103	315
NG133		0.3	23.7	11.3	254	1.1	3297	139	292
NG137		0.9	25.4	11.4	132	8.8	3237	127	284
NG195		0.7	30.0	10.4	30.6	0.1	3237	108	311
NG198		1.4	41.8	11.1	63	0.8	3237	77	292
NG220		0.8	45.4	12.0	17.6	0.5	3237	71	270
BL1		1.2	35.2	14.6	83	0.0	4376	124	300
BL18		3.0	75	17.5	119	4.1	3897	52	223
BL19		10.9	257	49.3	127	36.9	7134	27.8	145
BL28		0.9	44.4	16.6	114	3.6	5335	120	321
BL29		2.5	65	18.1	107	18.5	9232	142	510

Analyses performed at Leeds. Ti: measured as oxide with major elements.

Incompatible element concentrations and ratios are in general consistent with a model source slightly depleted in incompatible elements relative to a model primitive mantle compositions (Fig. 8.11). The Bend Formation samples are slightly richer in Ti, Zr and slightly poorer in Y than the Reliance Formation komatiites. The Ni contents of some Bend Formation komatiites are distinctly higher than Reliance Formation samples with equivalent MgO although there is considerable scatter in Ni contents. The most distinctive difference between the two suites is exhibited by the Ti/Y and Zr/Y trace element ratios where the relative enrichment in Ti and Zr compared with Y of the Bend Formation komatiites is apparent (Table 8.7, Fig. 8.11, Bend mean Ti/Y = 309 ± 22 (2 σ) whereas Reliance mean Ti/Y = 270±9). This difference cannot have been generated by contamination processes because crustal Ti/Y (ca. 270) is very similar to mantle Ti/Y (ca. 290) (Taylor & McLennan 1985). Instead, the difference probably reflects differences in source composition or in the petrogenesis of the lava suites.

The Brooklands Formation samples have compositions broadly comparable to the other suites.

205

8.7.2 *Contamination of magmas*

High temperature, low viscosity magnesium-rich komatiite liquids are thought to have flowed in a turbulent manner through the crust and after eruption (Nisbet 1982; Huppert et al. 1984). Turbulent flow would greatly enhance the ability of the magmas to transfer heat to wall rocks increasing the probability of their contamination by crust. Such contamination could cause substantial variation in the concentrations of those incompatible elements significantly enriched in continental crust.

Huppert & Sparks (1985a, b) calculate that komatiitic liquid with roughly 30% MgO could assimilate up to 30% of crustal rock during transport through a 40 km thick crust. The cooling effect of contamination and the resultant crystallization would drive the liquid composition into the basalt field. There are two important points relevant to the geochemistry of komatiitic lavas: 1) The most magnesian lavas are unlikely to be significantly contaminated unless derived from parent lavas of much more magnesian compositions; and 2) the most contaminated rocks will be basalt or komatiitic basalt in composition or olivine porphyries of komatiitic composition.

The more contaminated liquids will have crystallised as basalt or komatiitic basalt and Chauvel et al. in Chapter 7, argue that the range of initial $^{143}Nd/^{144}Nd$ isotopic compositions from komatiites, $\varepsilon_{Nd} = +3.2$, to komatiitic basalts with $\varepsilon_{Nd} = +0.4$ is consistent with formation of komatiitic basalt by fractionation and crustal contamination of komatiite. However, if the contaminant is ca. 3500 Ma crust as outcrops close to the SASKMAR 1 site, contamination is limited to 1% by weight of crust. The Pb isotopic compositions of the komatiitic basalts (Fig. 7.2, $\mu_1 = 8.4$) indicate a more enriched source (i.e. more old upper crustal contaminant) than those of the komatiites ($\mu_1 = 8.2$) as does the inverse correlation of Zr/Nb ratio with Zr (Fig. 8.11).

The extent of contamination of the komatiites is more difficult to assess as the original range of magma compositions is unknown. Their ε_{Nd} which ranges between ca. +2.5 and +3.2, and Pb isotopic compositions ($\mu_1 = 8.2$) are respectively lower and higher than the Archaean komatiites from the most depleted sources but such differences could result from 1% or less contamination by the nearby 3500 Ma crust. The element ratios which should be diagnostic of crustal contamination are those which exhibit most fractionation between crust and a chondritic primitive mantle (e.g. Taylor & McLennan 1985) or komatiite compositions. Such analysed element ratios include the REE (La/Yb 7.3 in average crust, 1.48 in chondrites and 0.88-1.01 in Reliance komatiites) and Zr/Ti ratios (0.0185 in average crust, 0.0085 in chondrites, 0.0079 ± 2 in Reliance komatiites, and 0.0088 ± 4 in Bend komatiite (Tables 8.7, 8.8). The range in La/Yb ratios (0.88 to 1.01), if due to contamination, would reflect only a variation between 0 to ca. 0.7% of crust added and the range of Ti/Zr ratios in Reliance Formation komatiites is only equivalent to a difference of ca. 1.5% contamination. If the komatiites are contaminated the contamination is remarkably uniform (close to analytical error on the ratios sensitive to contamination) and from the Pb and Nd isotopic data limited to ca. 1%.

At these levels the effect of contamination on the concentrations of the major elements and on all but the most incompatible elements will be close to undetectable, being for example only just significant for light rare earth elements and Zr or Sr. The concentrations of most incompatible elements, such as Rb and K, which would be substantially affected by contamination, are altered in komatiites but Pb with a ca. 60 fold enrichment in crust over primitive mantle would be substantially enriched. Crustal contamination of Pb is consistent with the heterogeneous Pb isotopic systematics of komatiites (Chauvel et al. 1983; Chapter 7) and the high μ_1 values of the Z-series samples.

The reason for lack of crustal contamination in the komatiitic lavas which were erupted onto continental crust and presumably passed through this crust is not known. Possibly the crust was very thin or eruption rates were significantly greater than estimates by Huppert et al. (1984) and Huppert & Sparks (1985a, b). Komatiites may have been transported upwards through dykes lined by the remains of previous mafic or ultramafic eruptions and the komatiites are erupted directly onto komatiitic flows. Similar processes occur at modern mid-oceanic ridges, where dykes appear to intrude preferentially through the centre line of the previous dyke which is the weakest part of the crust. In addition viscosity variation in non-Newtonian and cooler boundary layers might have inhibited convection at the margins of the dykes and flows substantially reducing their erosive capabilities below the theoretical estimates of Huppert and Sparks.

8.8 PETROGENESIS

8.8.1 *Komatiites*

The petrogenesis of komatiites has attracted much attention because of their significance as evidence for high temperatures in the Archaean mantle, the problems of generating high degrees of the melt at high levels in the mantle and the possible constraints placed by komatiites on mantle composition and structure.

The incompatible element concentrations and ratios in a wide range of komatiites are consistent with their derivation from a mantle with chondritic or slightly depleted trace element ratios at high degrees of melt and thus in equilibrium with olivine only, if at pressures less than ca. 50 kb (e.g. Bickle et al. 1976; Sun & Nesbitt 1976). In contrast the thermal problems of supplying the necessary latent heat of melting in addition to the high temperatures required (Bickle 1986), have led to proposals that komatiites represent low degrees of melt, very magnesian because of their derivation at very great depths. O'Hara et al. (1975) pointed out that initial melts become more magnesium rich with increasing pressure and also that many harzburgite nodules in kimberlite have an iron-magnesium ratio consistent with equilibration with komatiitic liquid. The possibility that harzburgite or a more complex garnet bearing mineralogy controls komatiite chemistry is therefore important.

The very coherent variation of the Reliance Formation komatiite compositions was early recognised as providing a possible test between olivine or harzburgite control on fractionation (Bickle et al. 1976; Nisbet et al. 1977). Small but apparently statistically significant deviations from olivine control were interpreted by Nisbet et al. (1977) as evidence for a more complex mantle petrogenesis (i.e. ol±opx rather than the olivine control expected at lower pressure). In particular it was noted that TiO_2-MgO variation passed outside olivine control (TiO_2 = 0 intercept of 47.2±3.6 MgO versus olivine compositions of 51% MgO) in contrast to the Al_2O_3-MgO variation. The major element variation in CMAS also deviated from olivine control. The problem of interpretation was that the deviations from olivine control were small and might have been caused by systematic alteration or contamination. The new data might be interpreted to support the previous interpretations. Zr-MgO variation indicates a similar intercept to TiO_2-MgO at 47% MgO (Fig. 8.10) and small systematic changes in Zr/Al_2O_3 and TiO_2/Al_2O_3 ratios as a function of MgO content are observed in the Reliance Formation samples. In contrast the Z-series samples discussed above preserve the chemical evidence for the observed in situ olivine fractionation although these samples are fresher than the other samples. Contamination is unlikely to alter TiO_2/Al_2O_3 ratios without altering Zr/Ti ratios as discussed above. The compositions of the lavas also deviate from olivine control when projected into CA-M-S (Fig. 8.12), a projection suggested by Herzberg & O'Hara (1985). However, the positions of points in this projection depend in CaO and SiO_2 concentrations and there is evidence that CaO concentrations are altered and the scatter

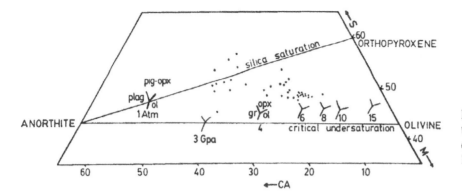

Figure 8.12. CA-M-S projection of komatiitic lavas from diopside after O'Hara (1968). Reliance Formation lavas.

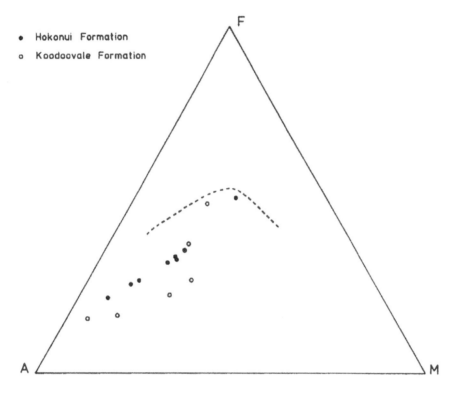

- Hokonui Formation
- Koodoovale Formation

Figure 8.13. AFM and Mg-Fe+Ti-Al cation plot of felsic volcanic rocks from Hokonui and Koodoovale Formations.

on SiO_2 concentrations is significant. The conclusion is that deviations from olivine control can be seen but that these are so small to be of uncertain significance. The significance of komatiite compositions to their source compositions and whether compositional differences between komatiite suites reflect source chemistry or their petrogenesis therefore remains uncertain and is likely to remain uncertain until the phase relations at the depths of probable komatiite generation are better known.

8.8.2 *Basalts and komatiitic basalts*

The basalts and komatiitic basalts exhibit much more scatter on variation diagrams. There is no obvious systematic difference between the Ngezi Group or Mtshingwe Group samples. Assessment of the effects of alteration and contamination on the basaltic rocks is much less easy than in the more coherent komatiite samples. In Chapter 7 the range of model initial ε_{Nd} values of between +3.2 for the komatiites to a minimum of +0.3

208

for komatiitic basalts is interpreted as evidence for contamination by up to about 1% of 3500 Ma continental crust in the komatiitic basalts. The range of incompatible element concentrations in the basalts cannot be explained by simple fractionation models (Fig. 8.9h), but it is not possible to distinguish the effects of source heterogeneity, melting processes or complexity of fractionation processes on their chemistry. The marked composition gap between komatiites and komatiitic basalts, such as is visible on Figure 8.9, has previously been interpreted as a result of density control on lava compositions erupted from komatiite-fed magma chambers (Nisbet & Chinner 1981). Some major mafic-ultramafic complexes such as the Shabani Complex may possibly be coeval with the Ngezi Group greenstones and might well represent remnants of the sort of magma chambers supplying both the Mtshingwe and Ngezi Group basaltic and komatiitic basalt lavas.

8.8.3 *The felsic volcanic rocks*

Analyses of felsic volcanic rocks from the Hokonui Formation and the felsic agglomerate member of the Koodoovale Formation are plotted on an AFM plot in Figure 8.13. These rocks plot in the calc-alkaline field, as is consistent with their andesitic chemistry (Table 8.7) and plagioclase-dominated modal compositions. These lavas, which are often pyroclastic, are relatively altered and slightly carbonated; this is undoubtedly reflected by the reset 2460 ± 100 Ma Rb-Sr isochron (see Table 2.2). Apart from noting their similarity to other Archaean calc-alkaline volcanic rocks (e.g. Hawkesworth & O'Nions 1977; Barley et al. 1984) little more can be said on the basis of the small suite studied.

8.9 CONCLUSIONS

The important conclusions from the Belingwe komatiite suites are:

1) Comparison of the very fresh with the more altered of the samples indicates that SiO_2, TiO_2, Al_2O_3, FeO, MgO, Ni, Cr_2O_3 and REE element concentrations are unaltered, but that some loss of CaO and substantial loss of Na_2O took place in more hydrated rocks. This lack of alteration is reflected in the preservation of olivine control on the chemical variation within individual layered flows and in the correspondence between rock chemistry and the compositions of phenocryst phases (mainly olivine).

2) The two suites of komatiites from stratigraphically distinct units are also compositionally distinct but the differences are slight and only readily apparent in trace element ratios (e.g. Ti/Y).

3) Crustal contamination in the komatiite lavas is probably less than 1% by mass unless all the komatiites in stratigraphically distinct suites are contaminated to precisely the same extent, an eventuality which is most unlikely.

4) The significance of variation trends in the Reliance Formation lavas, which deviate slightly from olivine control, is still uncertain.

5) The lowermost volcanic unit, the Hokonui Formation is a typical Archaean calc-alkaline sequence. This contrasts with the overlying formations which are dominated by basalt- komatiite lava.

APPENDIX

1 SAMPLE LOCALITIES AND DESCRIPTIONS

Z series all collected from 5.3 km south of Zvishavane (west of Zvishavane Golf Course – GR 81797456 from flows illustrated in Figure 8.2.

Z1 Lowermost in Joe's Flow. 250 cm below top. Plate olivine spinifex komatiite. Olivine 2-3 cm plates completely replaced fine yellowish pleochroic aggregate, pyroxene fresh, glass isotropic to altered, chromite euhedral with rare skeletal overgrowths; some talc and tremolite.

Z2 Joe's Flow 100 cm below top. Fine random spinifex. Very good preservation of 2-4 mm by 0.2 mm olivine plates only partly altered (Fig. 8.4). One area of polyhedral olivine rounded outlines 0.1-0.2 mm with skeletal overgrowths. Pyroxenes completely fresh, glass mostly isotropic. Chromite small euhedral to skeletal.

Z3 Joe's Flow, polyhedrally jointed flow top. Olivine grains with a variety of habits: 1) Polyhedral to slightly rounded 0.2-0.5 mm; 2) hopper olivine as overgrowths on (1) to skeletal grains and; 3) skeletal plates 0.3-0.5 mm long, 0.1 mm wide. Small fresh pyroxenes, minute chromite euhedra. More altered with only 10-30% of olivine fresh.

Z4 Tony's Flow, 100 cm above base, cumulate layer. Texturally similar to Z3 but much fresher and polyhedral olivines less abundant. Smaller skeletal plates partially serpentinised, all other grains largely fresh. 'Glass' devitrified possibly during quenching.

Z5 Tony's Flow, 500 cm above base. Olivine enriched cumulate. Olivine mostly hopper but some skeletal growths (Fig. 8.4), 0.05 to 1.0 to 25 mm for some elongate grains. Matrix coarse random pyroxene in isotopic glass. Very fresh olivine little serpentinised.

Z6 Tony's Flow, 430 cm below top. Cumulate layer. Polyhedral to rounded 0.01 to 3 mm olivine, rare skeletal overgrowths show optical discontinuity. Alteration only on margins and cracks. Pyroxenes nucleate on and mantle olivines, skeletal pyroxenes relatively coarse in fine granular dark coloured groundmass. Rare patches of carbonate plus talc in possible vesicles, small chromite euhedra.

Z7 Tony's Flow, 170 cm from top. Random spinifex B1 zone. Aligned polyhedral and minor hopper olivine, 5-10 mm by 0.2 mm with skeletal overgrowths. Olivine 10-30% serpentinised. Pyroxene acicular to frondlike and glass fresh. Chromite minute, ubiquitous.

Z8 Tony's Flow, 90 cm from top. Spinifex. Olivine elongate hopper (1-3 × 0.3 mm) to long fine wafers (1-2 ×0.1 mm). Hoppers earlier generation than wafers which overgrow hoppers. Olivine mostly serpentinised. Pyroxenes coarse, fresh spectacular twinning, glass isotropic, brown. Chromite coarse euhedral.

Z9 Massive Flow (spinifex free). Near base. Abundant coarse solid olivine, more hopper than plate. Not very fresh with probable weathering.

Z10 Massive Flow, 11 m below top (5 m above lowest polyhedral olivine (1.5-0.1 mm) few skeletal overgrowths minor, small hoppers. Pyroxene and glass matrix. Not too well preserved, olivine 50% unserpentinised with cloudy appearance.

Z11 Massive Flow, 5 m below top. Few polyhedral olivines (2 mm) in clusters, most olivine polyhedral (0.3 mm) with some skeletal overgrowths, some hopper. Fine grained matrix, euhedral chromite. Preservation better than Z10 but not good.

Z12 Massive Flow, polygonal jointed top. Similar to Z11, matrix coarser and rather altered with patches of massive, crustiform talc replacement.

210

2 BEND FORMATION KOMATIITES

Location	Description
SL413 GR 80507158	Small 0.5 mm polyhedral to rounded? cumulate olivine serpentinised. Matrix medium grained (0.1 mm) tremolite and opaques.
SL442 GR 80097160	Aligned olivine spinifex plates (10×0.5 mm) some skeletal, serpentinised. Matrix mainly very fine tremolite and opaque. Some fine skeletal pyroxene unaltered. Skeletal chromite.
SL445 GR 80097160	Fine random olivine spinifex (5×0.25 mm), some hopper/skeletal olivine, serpentinised, Matrix, brown very fine grained devitrified after glass, some coarser tremolite. Skeletal chromite.
SL501 GR 80237163	Random 1 cm olivine plate spinifex. Olivine and matrix recrystallised to fine aggregate of serpentine, chlorite, tremolite and opaques. Rather altered.
SL502 GR 80237163	Aligned 50×1 mm olivine plate spinifex. Olivine serpentinised and matrix mainly tremolite and opaques. Rather altered.
SL504 GR 80237163	Aligned 50×1 mm olivine plate spinifex, olivine replaced by serpentine and tremolite, matrix mainly tremolite, some chlorite, talc and opaques. Rather altered.
SL550 GR 80257194	Aligned coarse skeletal olivine spinifex plates (5 or more $\times$ 1 mm max.) completely serpentinised. Fine altered blades of olivine 2×0.1 mm in brown very fine grained distributed glass, partly replaced by tremolite in some areas.
SL585 GR 80497249	Olivine spinifex plates (10×0.5 mm) and numerous skeletal plates and hopper olivine, all serpentinised. Coarse (1×0.1 mm) acicular to stubby clinopyroxene unaltered in fine grained isotropic matrix. Skeletal chromite and opaques.
SL703 GR 80237163	Aligned 150×0.25 mm olivine spinifex plates or serpentinised in matrix of tremolite, opaques. Some tremolite pseudomorphs frond-like pyroxene.
SL708 GR 80357148	Felted mass of serpentine and opaque. No real igneous texture discernible. Very altered.
SL741 GR 80257194	Skeletal 0.5 mm hopper olivine and olivine platelets serpentinised. Matrix brown devitrified, pyroxene fronds partly replaced by tremolite, opaques, skeletal chromite.

3 BEND FORMATION KOMATIITIC BASALTS AND BASALTS

SL403 GR 80317160	Pillowed lava. Aggregate of tremolite, clinozoisite, quartz, sphene, rare chlorite and opaques. Tremolite probable pseudomorphic after clinopyroxene.
SL461	Clinopyroxene, hollow and skeletal acicular entirely replaced by tremolite and chlorite in hollow cores. Matrix tremolite, clinozoisite, sphene, quartz, opaques.
SL498 GR 80187159	Aggregate of tremolite (after chlorite-filled hollow clinopyroxene), clinozoisite, sphene and albite.
SL499 GR 80187159	Aggregate of tremolite, albite, sphene, chlorite.
SL526 GR 80467255	Foliated tremolite, epidote, albite, sphene, chlorite.
SL702 GR 80317160	Aggregate of tremolite, epidote, chlorite, albite, sphene. Tremolite pseudomorphs of skeletal acicular and hollow pyroxene.

SL713 GR 80467255	Pillowed lava. Spherulitic (light coloured spherical 1 cm felsic patches). Tremolite, epidote, chlorite, albite. Clinopyroxene as stubby laths partly altered to tremolite in finer matrix of tremolite, albite in felsic patches. Darker areas more altered and tremolite rich.
SL714 GR 80467255	Pillowed lava. Spherulitic texture. Hollow chlorite-filled clinopyroxene partly replaced tremolite in aggregate of tremolite, epidote, chlorite, albite, sphene and opaques.
SL715 GR 80467255	Pillowed lava. Spherulitic texture similar to SL714 more plagioclase and coarse epidote.
SL716 GR 80467255	Pillowed lava. Spherulitic texture, similar to SL715.
SL719 GR 80467245	Pillowed lava. Olivine as skeletal hopper and small (2 × 0.1 mm) plates pseudomorphed by chlorite. Matrix abundant plumose fronds of pyroxene replaced by tremolite, fine epidote and unresolved fine grained metamorphic minerals.
SL720 GR 80467245	Pillowed lava. Similar to SL719.
SL734 GR 79937148	Clinopyroxene spinifex. Fine random: 5 × 0.2 mm (max.) acicular and skeletal randomly oriented pyroxene pseudomorphed by tremolite with chlorite filling original hollows. Matrix tremolite, albite, epidote, sphene, chlorite.
SL736 GR 79937148	Clinopyroxene spinifex. Fine random, similar to SL734.
SL737 GR 79937148	Clinopyroxene spinifex. Fine random, similar to SL734.
SL739 GR 80257215	Pillowed lava. Spherulitic texture. Acicular skeletal and plumose sheaves of pyroxene pseudomorphed by tremolite, albite, epidote, sphene and chlorite in groundmass.
SL740 GR 80247215	Clinopyroxene spinifex. Slightly deformed. Sets of parallel needles (5 × 0.1 mm) replaced by tremolite in matrix of chlorite and fine grained alteration products.
SL745 GR 80227193	Pillowed lava. Spherulitic texture. 0.5 mm skeletal olivine hopper and plate crystals chloritised. Very fine grained brown groundmass with pyroxene 'fronds' replaced by tremolite.
SL752 GR 79817163	Pillowed lava. Similar to SL745 but slightly more recrystallised with coarser tremolite sheaves.
SL776 GR 80257215	Pillowed lava. Similar to SL745. Some epidote.
SL780 GR 80257215	Fine random olivine spinifex. 5 × 0.05 mm plates after olivine and some hopper olivine all chloritised. Matrix fine fronds of pyroxene replaced by tremolite.
SL781 GR 80257215	Fine random clinopyroxene spinifex. 5×0.1 mm skeletal pyroxene needles filled chlorite and altered to tremolite in matrix of tremolite, albite and epidote.
SL782 GR 80257215	Clinopyroxene spinifex. Similar to SL781.
SL785 GR 80257215	Clinopyroxene spinifex. Similar to SL740.

4 HOKONUI FORMATION FELSIC VOLCANIC ROCKS

GC11 GR 80457305	Pyroclastic volcanic 0.5 cm angular clasts of tremolite, epidote and albite (fine grained) in quartz-rich, chlorite bearing matrix. Possibly silicified.

GC18 GR 80517227   Pyroclastic volcanic rock. Very fine grained dark clasts in matrix (foliated) of fine grained quartz sericite, carbonate rhombs (about 1%). Also numerous partly altered 0.5 mm albite porphyroclasts.

5 KOODOOVALE FELSIC VOLCANIC ROCKS

SL728 GR 807 717   Felsic agglomerate. Numerous 0.5 mm albite porphyroclasts possibly phenocrysts, euhedral, partly altered to carbonate, sericite and epidote in matrix of fine grained albite laths, sericite, epidote and some carbonate.

SL729 GR 807 717   Felsic agglomerate, 0.5 mm corroded and rounded quartz phenocrysts, very altered (to fine grained aggregates) plagioclase phenocrysts in fine grained matrix containing epidote, carbonate and opaques.

SL730 GR 807 717   Felsic agglomerate. Similar to SL729 except fewer quartz phenocrysts, plagioclase less altered but to coarser grained sericite, quartz, plagioclase, sericite, chlorite and carbonate in groundmass.

SL731 GR 807 717   Felsic agglomerate. Similar to SL730 but fewer plagioclase phenocrysts.

SL733 GR 807 717   Felsic agglomerate. Similar to SL729.

6 BROOKLANDS FORMATION KOMATIITES, KOMATIITIC BASALTS AND BASALTS

NG728 GR 970267   0.5-1 m amygdaloidal pillows, rather deformed. Texture rather recrystallised but probably after skeletal hopper and 0.5×0.1 mm plate olivine now chloritised in matrix. Frond pyroxene recrystallised to tremolite in matrix of tremolite, chlorite, albite, quartz, epidote and opaques.

NG730 GR 970267   Pillow lava. Similar to NG728.

NG731 GR 970267   Pillow lava. Similar to NG728.

NG771 GR 948245   Columnar clinopyroxene spinifex (30 cm zone) in 7 m flow with olivine cumulate base. Skeletal acicular pyroxene replaced by tremolite and filled by chlorite. Matrix tremolite, chlorite, albite.

NG772 GR 948245   Columnar clinopyroxene spinifex zone from similar flow below that of NG771. Similar to NG771.

NG775 GR 948245   Massive lava flow or sill. Aggregate of actinolite (pale blue green pleochroic), albite and opaques.

NG776 GR 952246   Deeper in same flow as NG775. Similar mineralogy.

NG777 GR 952246   Collected below NG776 in same flow as NG775. Similar mineralogy.

NG780 GR 952246   Massive lava flow or sill. Similar to NG775.

CHAPTER 9

Controls on the formation of the Belingwe Greenstone Belt, Zimbabwe

E.G. NISBET, M.J. BICKLE, J.L. ORPEN & A. MARTIN

ABSTRACT

The Belingwe Greenstone Belt, with its excellent stratigraphic preservation, unconformities, komatiites and stromatolites, provides information on the structural, sedimentological, volcanological and palaeontological evolution of some Archaean rock
suites. Nevertheless a number of geological problems remain unresolved.

The basement to the belt includes an ancient granitoid and gneiss terrain, of uncertain
heritage prior to 3500 Ma and granitoid and gneissic rocks formed about 2900 Ma. On
this was laid down the Hokonui succession, which may represent a single large andesitic
volcano. Above and to the side of the Hokonui succession, subsidence allowed deposition of the alternating suite of komatiitic volcanic rocks and ironstones of the Bend
Formation. This episode was terminated by marginal uplift and deposition of the conglomerates and felsic volcanics of the Koodoovale Formation. Possibly at the same time,
the Brooklands Formation was deposited 60 km or more to the east in an actively faulted
graben.

The Ngezi Group was probably formed in a later extensional event, in which initial
sedimentation was followed by komatiitic volcanism (briefly) and then a substantial
period of basaltic volcanism. Rift phase sedimentary deposits associated with this event
have not been identified. It is possible that some of the mafic volcanics were derived from
a mid-crustal magma chamber similar to those preserved in the large ultramafic complexes of central Zimbabwe.

The structural evolution of the belt may have been consequent upon its initial rifting
history. Offlapping strata may have been tightened into a syncline in which the preserved
structural thickness is substantially increased by stacking of laterally juxtaposed depositional facies. Whether or not the site of the syncline reflects an initial basin within the
craton-wide upper Bulawayan sequence remains controversial.

9.1 INTRODUCTION

Little is known about the tectonic environment(s) in which greenstone belts formed,
although there are many theories. The Archaean is the last refuge of that endangered
species, the 'tectogene,' and much has been written about 'downsagging' in basins,
produced by a supercargo of volcanics on a weakened crust. All these ideas founder on

215

the rock of isostasy: The processes cannot be self-perpetuating as the less dense infra-structure of volcanic and sedimentary rocks must subside by replacing more dense mantle. In the Belingwe Belt the available information as to 'how' the belt developed is more detailed than in most greenstone belts; perhaps this detailed information may provide some clue as to the tectonic causes of subsidence and the associated igneous activity. In the summary below we are still concerned principally with questions related to 'how' not 'why.' Direct observation cannot be used to reveal the 'scheme of Archaean tectonics' in the way that plate tectonics has been proven, but an appraisal of appropriate tectonic models may raise previously unasked and important questions concerning field relations.

9.2 WHAT HAPPENED: THE BASEMENT

The history of the 3500 Ma Shabani Gneiss basement and of the 2900 Ma Chingezi Gneiss basement is shrouded in mystery. The siting of the Belingwe Greenstone Belt on the junction of these two terrains is of unknown significance. Speculations on the origin of greenstone belt-related gneissic terrains, and their possible relationship to horizontal (overthrust) high-grade gneiss belts are probably best answered elsewhere (e.g. in Western Australia, Bickle et al. 1985).

9.3 THE MTSHINGWE GROUP GREENSTONES

The older of the two major greenstone groups which comprise the Belingwe Greenstone Belt provides some interesting information on Archaean tectonics although there are problems in understanding the structural setting and correlation of the component forma-tions. The Hokonui Formation is an andesitic volcanic pile, with dimensions similar to other Archaean volcanic centres and to large modern andesitic volcanoes. This volcano may have grown on a gneissic crust that was only slightly older than itself. The volcanic pile was intruded by what may have been an essentially synchronous cogenetic tonalite. Such Archaean volcanic-plutonic calc-alkaline provinces are petrologically and geo-chemically very similar to modern calc-alkaline volcanic-plutonic belts situated along continental margins above subduction zones (e.g. Bickle et al. 1983; in press.). The 2950 to 3050 Ma model T_{DM} Sm-Nd ages on the associated Chingezi Tonalite (Taylor et al. 1991) indicate that the '2900 Ma' event involved the derivation of significant volumes of crust ultimately from mantle sources at that time. It is possible that the widespread felsic volcanic correlatives of the Hokonui Formation found across the western Zimbabwe craton (Wilson et al. 1978; Wilson 1979; Wilson et al. 1990) are remnants of a line of stratovolcanoes above an Archaean subduction regime in a continental margin setting.

The Hokonui Formation was overlain by a sequence of alternating komatiite, mafic volcanic, chert and ironstone constituting the Bend Formation. The thickness of this formation probably averages about 3 km, although exposure and structural relations are poor and it is not possible to be certain that tectonic repetition or transport of this unit has not occurred. The overlying conglomerate-dominated Koodoovale Formation is of par-ticular significance. It has a well exposed channelled base on the Bend Formation and includes clasts with composition typical of nearly all the underlying rocks: Bend Forma-tion, Hokonui Formation, intrusive adamellite and (significantly) clasts similar to the Chingezi Gneiss, which must have been basement by then.

The Brooklands Formation is correlated with the other Mtshingwe Group green-

216

stones, but only on the basis of structural setting. It is thought to have been deposited on a basement of ca. 3500 Ma Shabani Gneiss. The evidence for this unconformity is strong but indirect. It includes the structural and metamorphic contrast between the Brooklands Formation and the basement, a mapped but not exposed unconformity, and the apparent derivation of clastic material from a proximal gneissic terrain. The most important conclusions from study of the Brooklands Formation are that its sedimentology and outcrop distribution are consistent with deposition in a deepening fault bounded graben. Its dimensions (Fig. 5.2) are similar to modern grabens in rifted continental crust (Jackson & McKenzie 1983) as illustrated in Figure 9.1, with some coarse proximal fan deposits as well as deeper water facies. Reversal of motion on this fault during the compressional phase which synclinally folded the Mtshingwe Greenstones may have uplifted this facies, which has then been partially removed by erosion. The more northerly remnants of the Brooklands Formation (Fig. 2.1) might represent fragments of separate structures, given their 30-40 km separation.

Some important conclusions which can be drawn from the Mtshingwe Group greenstones are 1) the variability of stratigraphic sequences; 2) the variable setting of komatiite volcanics, which occur above felsic volcanic rocks (e.g. Bend Formation) or within deeper water sediments but above shallow water facies (Brooklands Formation); and 3) the inference that the deposition of the Brooklands Formation occurred in an active rifted continental tectonic environment on the same scale as modern extensional rifts.

9.4 THE NGEZI GROUP GREENSTONES

9.4.1 *The stratigraphic setting*

There are well developed basal unconformities exposed between the shallow water Manjeri Formation sedimentary rocks and the underlying ca. 3500 Ma gneissic basement, the Bend Formation and the Hokonui Formation. There is a mapped unconformity between the Manjeri and the Brooklands Formation. Collectively this evidence unequivocally demonstrates that the Manjeri Formation was laid down on an older granite-greenstone crust. The very continuous overlying stratigraphic succession contains rock types typical of Archaean greenstone belts (komatiitic and basaltic lavas and sediments) and is plausibly interpreted as a conformable succession. These relationships demonstrate that this particular greenstone belt sequence was formed above continental crust.

A number of other unconformities between greenstone belt sequences and older continental crust have been reported. In the Mashava area 80 km east of the Belingwe Belt sediments correlative with Manjeri exhibit a well exposed unconformity over the 2950 Ma Mushandike Granite (Orpen & Wilson 1981; Taylor et al. 1991). In the Slave structural province, Canada, several unconformities between sediment and granitoid are well described (Ermanovics & Davison 1976; Henderson 1975) but the 'greenstone' stratigraphy is unusual as komatiites are absent. At Steep Rock in the Wabigoon Greenstone terrain, Ontario, a well exposed basal unconformity is preserved. Here, a complex granitoid gneissic basement is overlain by basal grits and conglomerates, upon which is a thick reef of stromatolitic limestone, a major iron ore unit and volcanoclastic rocks including high-magnesian eruptions (Jolliffe 1966; Wilks & Nisbet 1985, 1988). Structural complexity in this area obscures the nature of the succession above the volcanoclastic rocks. In the Yilgarn Block, Western Australia, Durney (1972) reported an unconformity between the Jones Creek conglomerate and a granitic basement but the stratigraphic significance of this unconformity is controversial (e.g. Naldrett & Turner 1977).

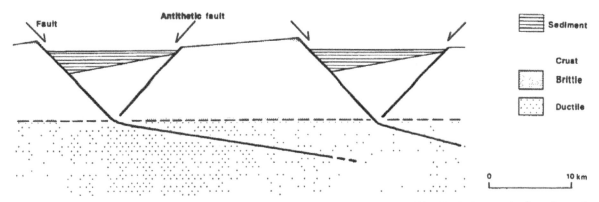

Figure 9.1. Illustration of the geometry of an active graben after Jackson & McKenzie (1983). Listric faults at depth conjectural. Depth to base of brittle zone depends on crustal thermal gradients.

Unconformable contacts between greenstone sequences and granitoid basement have also been described or inferred in India (Holland 1902; Chadwick et al. 1978) and Tanzania (Haidutov 1976). Evidence of major unconformities within greenstone belts are numerous in the literature. In some, komatiite-dominated sequences overlie older felsic volcanic sequences. There are also many examples of granitic clasts occurring in sedimentary sequences interbedded within greenstone belt successions. Evidence of shallow water sedimentation or even evaporitic deposits within greenstone volcanic sequences is also widespread (Barley et al. 1979; Lowe & Knauth 1977; Buick & Dunlop 1990). Additional evidence for older continental basement under komatiite-bearing sequences is provided by older xenocryst zircons (e.g. Compston et al. 1986; Thorpe et al., in press) and systematic contamination of Sm-Nd isotopic systematics (Chauvel et al. 1985).

The formation of greenstone belts above continental crust appears to have been a widespread phenomenon and it is perhaps the complexity of the subsequent tectonic and plutonic histories of the belts that has made the basal relations to hard to decipher. It is important to note that the present tectonic complexity of many greenstone belts does not necessarily imply an oceanic origin for greenstones as implied for example by Burke et al. (1976). Conversely, the demonstration here that this greenstone belt is of supra-continental origin does not imply that all Archaean mafic volcanic rocks were formed in the same setting. Each area must be interpreted on its own evidence.

9.4.2 Basin formation and tectonics

An important question concerning the nature of the Belingwe Greenstone Belt is whether the greenstone belt was deposited in an isolated local basin centred on the present synclinal fold or whether the fold preserved part of a more widespread sequence. This question has not been answered. The correlation of the Belingwe stratigraphy widely across the central and southern parts of the craton (Wilson et al. 1978; Wilson 1979) suggests a widespread cover but that does not preclude significant thickness variation (Nisbet et al. 1981). Sedimentological evidence of facies variation and current directions in relation to a localised basin is ambiguous although it is intriguing that current directions are mainly along the present axis of the belt. There is however, no direct evidence of a localised basin but the exposed section of the Manjeri Formation is perhaps too limited to provide it. The facies variation in the Cheshire Formation is distinctly asymmetric with respect to the

218

folding. Furthermore, the clastic component of the Cheshire Formation is predominantly volcanic-derived with minor arkose and quartzite and no granitic clasts: the granitoid crust was probably not exposed at the Cheshire stage of greenstone belt evolution. The only evidence consistent with localised deposition is that Manjeri sediments are absent in the higher-grade northern parts of the belt where deeper, central structural levels are exposed. This could be interpreted as representing the rapid attainment of deeper water conditions in the more central parts of the greenstone belt with little or no sedimentation before volcanism. Poor exposure, the presence of intrusive granites and higher strain in these areas precludes confidence in this interpretation.

Understanding the tectonic setting of the depositional basin poses major problems. What was the cause of the 5 to 10 km subsidence needed to accommodate the greenstone belt? The succession from shallow water sedimentary facies in the Manjeri Formation, to mainly pillowed subaqueous volcanicity, followed by shallow water sedimentation in the Cheshire Formation attests to the total magnitude of subsidence. The small proportion of resedimented volcanic detritus within the volcanic formations demonstrates the rapidity of this subsidence during the volcanic phase. It is possible that the stacking of units with lateral variations of thickness during the folding of the greenstone belt has increased the apparent stratigraphic thickness (Fig. 9.4) but even so several kilometres of subsidence must be required. The problem is a familiar one in the geological record: How did the sedimentary basin form? In general, a number of causes of sedimentary basin formation undoubtedly exist. These include the flexural response to the loading of crust by thrust sheets (e.g. Beaumont 1981), crustal extension (McKenzie 1978), and also (more speculatively) processes such as intrusion of mafic material and/or phase changes in the deep crust and in the lithospheric upper mantle (see review by Fowler 1990). The lack of voluminous clastic detritus in the Ngezi Group, as would be expected adjacent to an emergent mountain belt, rules out a foreland basin environment and it seems most probable that the subsidence was a result of extension. Rapid lithospheric extension with consequent adiabatic decompression of the underlying mantle provides a mechanism for the associated volcanism (McKenzie & Bickle 1988).

McKenzie et al. (1980) and Bickle & Eriksson (1982) suggested that an extensional model is widely applicable to Archaean greenstone belts. Bickle & Eriksson (1982) emphasised that the rapid 'initial rifting' or 'proto-basinal' phase of subsidence, when crustal extension thins the crust (but the isostatic subsidence is partly offset by the upwelling of hot less dense asthenosphere) should be clearly distinguishable from the subsequent much slower thermal phase of subsidence when the subsidence is caused by the increase in sub-crustal density during slow conductive relaxation of the thermal state of the lithosphere towards equilibrium. Thermal subsidence is significant for about 200 Ma in recent sedimentary basins (e.g. Wood & Barton 1983). The volcanic-dominated lower part of greenstone sequences was thought by Bickle & Eriksson (1982) to be a consequence of the rifting phase with overlying sedimentation characteristic of the thermal phase.

The facies sequence in the Manjeri Formation, which shows a transition from shallow to deeper water facies, followed by the copious volcanicity of the Reliance and Zeederbergs Formations from the hotter Archaean mantle is consistent with such a scenario, although the strata are very thin. However the associated extension would be expected to produce rifts of the dimensions shown in Figure 9.1. The protobasinal stage should be better developed. Possibly it is recorded as the underlying Brooklands Formation though the apparent time gap at the unconformity (ca. 150 Ma, Chapter 2) and the post-Mtshingwe Group erosional event make this unlikely. More likely, the pre-Manjeri surface was mature and contributed little detritus to a rapidly subsiding basement. In

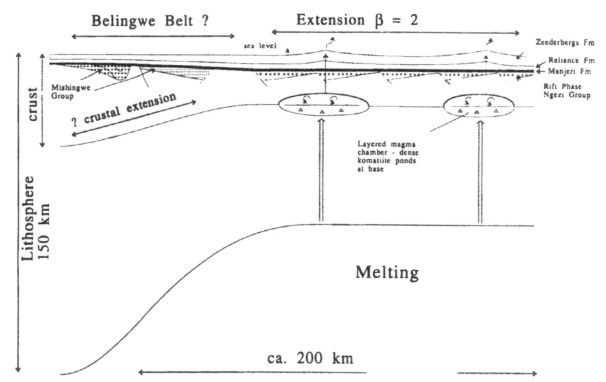

Figure 9.2. Possible model for development of the Reliance and Zeederbergs Formations. Note vertical exaggeration of Manjeri and Reliance Formations. Belingwe Greenstone Belt situated adjacent to actively extending upper crust. Precise correlation of Manjeri Formation and volcanics with rift-fill facies is unknown. After initial Manjeri-subsidence, melts from the asthenosphere are able to rise directly to the surface. Later, crustal magma chambers develop and only light fractionated basaltic liquid can ascend. Residue of doubly-diffusive magma chamber remains as crustal ultramafic complex.

either case, major rifts should be recorded in the Manjeri Formation. Although there is some stratigraphic and thickness variation within the Manjeri Formation such rifts are not apparent in the Ngezi Group stratigraphy. Faulting may have been parallel to present outcrop strike and not exposed, the rifts may have been situated on areas now occupied by granitoid (perhaps a consequence of rifting) or rifting may have taken place off the presently exposed margins of the craton. Figure 9.2 illustrates such a scenario with the Belingwe Greenstone Belt situated marginal to rifted crust. The subsidence is either due to non-uniform stretching of crust or possibly the elastic flexural response to loading of the lithosphere (e.g. Watts et al. 1982). The uniform thickness and wide extent of the volcanic sequences is to be expected of high-temperature low-viscosity lavas. An illustration of this is the ca. $10^6 km^2$ extent of the Mesozoic Deccan Trap flood basalts (Cox 1989). Further speculation would require better constraints on regional facies variation, tectonic setting and the timing of events.

9.4.3 *The derivation and petrogenesis of the lavas*

The wider aspects of the genesis of komatiites have been discussed in detail elsewhere (e.g. Arndt & Nisbet 1982; Nisbet 1987), with much attention being paid to lavas from Belingwe. A topic which has not been much discussed, however, is the relationship between komatiites, komatiitic basalts and basalts. In Chapters 7 and 8 a case was made that some of the lavas had been contaminated on ascent while others had not. The

220

komatiites had undergone at most addition of 1% of crustal material whereas some komatiitic basalts had assimilated a few percent of crust. Those lavas which did arrive at the surface, contaminated or uncontaminated, form a markedly bimodal suite, including a relatively small component of very magnesian lavas in the Reliance Formation, erupted soon after but not at the beginning of the volcanic event, and a much larger and diverse component of less magnesian lavas which include both the initial lavas of the succession and also the great monotonous bulk of the Zeederbergs Formation.

Figure 9.2 shows a possible model for the tectonic evolution of the belt in Reliance-Zeederbergs time, and for the petrogenetic relationships between the various lava components. In the earliest stages of the development of the Ngezi Group, a regime of regional extension produced rapid subsidence and thinning of continental lithosphere (probably ≥ 150 km thick initially, Bickle 1986), with attendant uprising of underlying asthenosphere. Pressure-release melting produced the volcanism, with the komatiites possibly generated in an anomalously high temperature mantle plume at great depths (Campbell et al. 1989). Calculations using the methods of McKenzie & Bickle (1988), but with a modified peridotite solidus after Herzberg et al. (1990), indicate that ca. 12 km melt would be produced by $\beta = 2$ extension of a 150 km thick lithosphere overlying mantle with a potential temperature of 1600°C, and 25 km of melt above a mantle of 1700°C. Average melt compositions would be komatiitic basalt. The low-viscosity lavas would be likely to flow well outside the rifted region reducing the preserved lava thickness. The most magnesian komatiites with up to 27 to 30% MgO were erupted early after minor initial komatiitic basalt volcanism. The komatiites were followed by the voluminous komatiitic basalt and basalt eruptions. It is probable that later komatiites became trapped at the density change at the base of the crust or where, at intermediate levels, the crust makes the change from dense, garnetiferous deep crustal assemblages to lighter granitoids. Here, slowly ascending dense komatiitic liquids would pond, develop magma chambers and interact with the enveloping continental crust. In such a setting, incoming komatiitic melt would have fractionated to evolved minimum density and variably contaminated basaltic liquids (Fig. 9.3). In a steady state magma chamber, parental liquid would have never reached the top of the chamber. Instead, only basalt would have been erupted (Nisbet & Chinner 1981). Below the basaltic liquid, in a layered, doubly-diffusive chamber, ultramafic cumulates would have collected. It is possible that some of the lavas of the Zeederbergs Formation were erupted from such a setting, and large layered complexes such as the Shabani and Mashava complexes may represent examples of magma chambers exhumed from intermediate crustal levels. However incompatible element enrichments in many of the basalts and komatiitic basalts (e.g. Zr, Fig. 8.8d) indicate that most of the less magnesian lavas were derived as smaller degree melts of the mantle and not as fractionates of a komatiitic parent.

Why the erupted komatiites escaped contamination is problematic. Possibly the first erupted komatiitic basalts represent such contaminated komatiites and these rocks both warmed and lined the dyke-like conduits up which the subsequent komatiites were transported in a setting thereby protected from the thermal and chemical effects of the crust. On eruption, the lavas would have poured out over earlier komatiite.

9.5 STRUCTURAL EVOLUTION OF THE GREENSTONE BELT

221

The structural interpretation of the Belingwe Greenstone Belt is limited to discussion of the geometry and relationships of component units. The structural interpretation of the Mtshingwe Group greenstones is consistent with refolding about the main synclinal axis

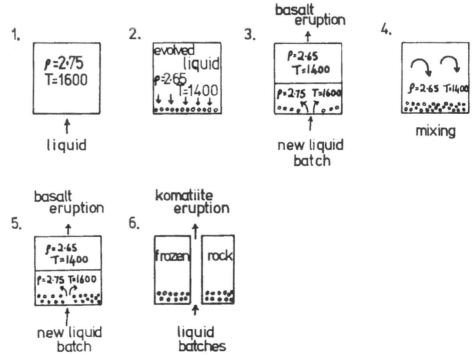

Figure 9.3. Detail of processes operating in magma chamber of Figure 9.2. Dense komatiitic parent liquid rises into crust. In regime of high tension, it can ascend to surface directly. Some such melts (especially early melts) may be contaminated by crust on ascent; other melts may ascend by injecting through the weakest section of the lithosphere, which is the centre of the previous dykes, and thus reach the surface with little contamination. At a later stage, with lower tension, more slowly ascending melts are trapped by density contrast in the crust and accumulate to form a magma chamber. Here a light fractionate will be produced as olivine precipitates, and wall-rock contamination may also occur. Further injection of melt from below would form a layered doubly diffusive chamber, and only light basaltic liquids would be able to erupt. Only in the event that the magma chamber froze over completely and significant regional tension built up again would it be possible for komatiitic liquids to ascend directly to the surface.

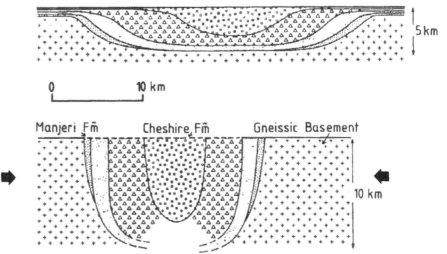

Figure 9.4. Illustration of stacking of lateral thickness variations to increase apparent stratigraphic section.

of the Ngezi Group greenstones. The simplest interpretation is that prior to deposition of the Ngezi Group the Mtshingwe Group greenstones were preserved in simple synclinal folds and minor partially-eroded anticlinal folds. The average thicknesses of the Mtshingwe greenstone sequence would be about 5 km but would have reached a maximum of 10 km in the Bend syncline if the shape of present outcrop represents true cross-sections.

In contrast, a possible present-day shape for the Ngezi Group Greenstone Belt is illustrated in Figure 3.18. This model is based on the hypothesis that the higher-grade northerly outcrop represents a greater uplift (and thus a deeper level exposure) of a 'pinched' foliated closure. Such a structure is consistent with relatively shallow (5-8 km) depths of greenstone implied by gravity data (Swain & Podmore, pers. comm.). The shape is also consistent either with a compressional model in which folding of the greenstones generated dome and cusp structures, or with a diapiric model, as discussed in Chapter 3. The steep dips at the surface are not explained by either of these models (but note the shallower dips in the possibly coeval Shabani ultramafic complex). The model does not address the problem of the extent of Chilimanzi Suite intrusion at depth. The cross-cutting relationships between the Chilimanzi Suite and the first folding event suggest that the formation of the greenstones and folding to their present synclinal form took less than 120 Ma (2692 ± 9 Ma to 2570 ± 25 Ma, see Chapter 2). Subsequent tectonic events, the Chilimanzi Suite intrusions and the D_g3 cleavage-forming event and subsequent faulting events were of regional scale across the craton and in part relate to major events within the Limpopo Belt.

9.6 CONCLUSIONS

The most important stratigraphic conclusion from this study of the Belingwe Greenstone Belt is that the Ngezi Group with its well-defined sedimentary and volcanic stratigraphy including komatiites was deposited on an older granite-greenstone continental crust. The details of the Belingwe stratigraphy have lead to a major revision of the Archaean stratigraphy of Zimbabwe set up by Macgregor (1951). One result of this revision is the widely recognised correlation (Wilson 1979) of the Ngezi Group and Mtshingwe Group with similar greenstones across the central and southern parts of the craton.

The field work has resulted in discoveries both of extrusive komatiitic rocks and of stromatolitic limestones which have been considerable significance to fields outside the scope of this work. In part, this volume is intended to serve as a substrate for those investigations. The structural and sedimentological observations place important constraints on models for the formation and evolution of the greenstone belt. The results indicate the necessity of detailed studies in Archaean granite- greenstone terrains and it is likely that much more information is yet to be gained from further sedimentological, volcanological, structural, metamorphic and geochronological work in the Belingwe Belt and surrounding area.

References

Abell, P.I., J. McClory, A. Martin & E.G. Nisbet 1985a. Archaean stromatolites from the Ngesi Group, Belingwe Greenstone Belt, Zimbabwe; preservation and stable isotopes-preliminary results. *Precamb. Res.* 27: 357-383.

Abell, P.I., J. McClory, A. Martin, E.G. Nisbet & T.K. Kyser 1985b. Petrography and stable isotope ratios from Archaean stromatolites, Mushandike Formation, Zimbabwe. *Precamb. Res.* 27: 385-398.

Anhaeusser, C.R. 1979. The evolution of the early Precambrian crust of southern Africa. *Phil. Trans. Roy. Soc. Lond.* A273: 359-388.

Arndt, N.T. 1977. Partitioning of nickel between olivine and ultrabasic and basic komatiite liquids. *Carnegie Inst. Wash. Ybk.* 76: 553-557.

Arndt, N.T. 1986. Differentiation of komatiite flows. *J. Petrol.* 27: 279-301.

Arndt, N.T. & M.E. Fleet 1979. Stable and metastable pyroxene crystallisation in layered komatiitic lava flows. *American Mineralogist* 64: 856-864.

Arndt, N.T., A. Naldrett & D.R. Pyke 1977. Komatiitic and iron-rich tholeiitic lavas of Munro Township, northeast Ontario. *J. Petrol.* 18: 319-369.

Arndt, N.T. & R.W. Nesbitt 1982. Geochemistry of Munro Township basalts. In N.T. Arndt & E.G. Nisbet (eds), *Komatiites*. London: Allen & Unwin, 309-330.

Arndt, N.T. & E.G. Nisbet (eds) 1982a. *Komatiites*. London: Allen & Unwin, 526 pp.

Arndt, N.T. & E.G. Nisbet 1982b. What is a komatiite? In N.T. Arndt & E.G. Nisbet (eds), *Komatiites*. London: Allen & Unwin, 19-27.

Ayres, C.D., P.C. Thurston, K.D. Card & W. Weber 1985. Evolution of Archaean supracrustal sequences. *Geol. Assoc. Canada Spec. Publ.* 28: 380.

Barley, M.E., J.S.R. Dunlop, J.E. Glover & D.I. Groves 1979. Sedimentary evidence for an Archaean shallow-water volcanic-sedimentary facies, eastern Pilbara Block, Western Australia. *Earth Planet. Sci. Lett.* 43: 74-84.

Barley, M.E., G.C. Sylvester, D.I. Groves, G.D. Barley & N. Rogers 1984. Archaean calc-alkaline volcanism in the Pilbara block, Australia. *Precamb. Res.* 24: 285-321.

Barton, J.M. & R.M. Key 1981. The tectonic development of the Limpopo Mobile Belt and the evolution of the Archaean cratons of Southern Africa. In A. Kroner (ed.), *Precambrian Plate Tectonics*. Amsterdam: Elsevier, 185-212.

Beaumont, C. 1981. Foreland basins. *J. Geophys. Soc.* 65: 291-329.

Beswick, A.E. 1982. Some geochemical aspects of alteration and genetic relations in komatiitic suites. In N.T. Arndt & E.G. Nisbet (eds), *Komatiites*. London: Allen & Unwin, 283-308.

Bickle, M.J. 1978. Heat loss from the earth: A constraint on Archaean tectonics from the relation between geothermal gradients and the rate of plate production. *Earth Planet. Sci. Lett.* 40: 301-315.

Bickle, M.J. 1982. The magnesium contents of komatiitic liquids. In N.T. Arndt & E.G. Nisbet (eds), *Komatiites*. London: Allen & Unwin, 477-494.

Bickle, M.J. 1986. Implications of melting for stabilisation of the lithosphere and heat loss in the Archaean. *Earth Planet. Sci. Lett.* 80: 314-324.

Bickle, M.J., L.F. Bettenay, M.E. Barley, D.I. Groves, H.J. Chapman, I.H. Campbell & J.R. de Laeter 1983b. A 3500 Ma plutonic and volcanic calc-alkaline province in the Archaean east Pilbara block. *Contrib. Mineral. Petrol.* 84: 25-35.

Bickle, M.J., L.F. Bettenay, C.A. Boulter, D.I. Groves & P. Morant 1980. Horizontal tectonic interaction of an Archaean gneiss belt and greenstones, Pilbara Block, Western Australia. *Geology* 8: 525-529.

Bickle, M.J., L.F. Bettenay, H.J. Chapman, D.I. Groves, N.J. McNaughton, I.H. Campbell & J.R. de Laeter (in press). Origin of the 3500 Ma-3300 Ma calc-alkaline rocks in the Pilbara Archaean: Isotopic and geochemical constraints. *Precamb. Res.*

Bickle, M.J., H.J. Chapman, L.F. Bettenay & D.I. Groves 1983a. Lead ages, reset rubidium-strontium ages and implications for the Archaean crustal evolution of the Diemals area, Central Yilgarn Block, Western Australia. *Geochim. Cosmochim. Acta* 47: 907-914.

Bickle, M.J. & K.A. Eriksson 1982. Evolution and subsidence of early Precambrian sedimentary basins. *Phil. Trans. Roy. Soc. Lond.* A305: 225-247.

Bickle, M.J., C.E. Ford & E.G. Nisbet 1977. The petrogenesis of peridotitic komatiites: Evidence from high-pressure melting experiments. *Earth Planet. Sci. Lett.* 37: 97-106.

Bickle, M.J., C.J. Hawkesworth, A. Martin, E.G. Nisbet & R.K. O'Nions 1976. Mantle composition derived from the chemistry of ultramafic lavas. *Nature* 263: 577-580.

Bickle, M.J., A. Martin & E.G. Nisbet 1975. Basaltic and peridotitic komatiites, stromatolites and a basal unconformity in the Belingwe Greenstone Belt, Rhodesia. *Earth Planet. Sci. Lett.* 27: 155-162.

Bickle, M.J., P. Morant, L.F. Bettenay, C.A. Boulter, T.S. Blake & D.I. Groves 1985. Archaean tectonics of the Shaw Batholith, Pilbara Block, Western Australia: Structural and metamorphic tests of the batholith concept. *Geol. Assoc. Canada Spec. Publ.* 28: 325-341.

Bickle, M.J. & E.G. Nisbet 1977. The chicken or the egg. *Geology* 5: 68.

Bickle, M.J. & E.G. Nisbet (this volume). Sedimentology of the Brooklands Formation, Zimbabwe: Development of an Archaean greenstone belt in a rapidly-rifted graben. In M.J. Bickle & E.G. Nisbet (eds), *The Geology of the Belingwe Greenstone Belt: A Study of the Evolution of Archaean Continental Crust.* Geol Soc. Zimbabwe Spec. Publ. 2. Rotterdam: Balkema.

Blatt, H., G. Middleton & R. Murray 1972. *Origin of Sedimentary Rocks.* Englewood Cliffs, N.J.: Prentice Hall, 634 pp.

Bliss, N.W. & P.A. Stidolph 1969. A review of the Rhodesian Basement Complex. *Spec. Publ. Geol. Soc. S. Africa* 2: 305-333.

Buick, R. & R.S.J. Dunlop 1990. Evaporitic sediments of early Archaean age from the Warrawoona Group, North Pole, Western Australia. *Sedimentology* 37: 247-277.

Burke, K., J.F. Dewey & W.S.F. Kidd 1976. Dominance of horizontal movements, arc and microcontinental collisions during the later permobile regime. In B.F. Windley (ed.), *The Early History of the Earth.* London: Wiley, 113-129.

Cameron, W.E. & E.G. Nisbet 1982. Possible Phanerozoic analogues of komatiitic basalts and komatiites. In N.T. Arndt & E.G. Nisbet (eds), *Komatiites.* London: Allen & Unwin, 29-50.

Campbell, I.H., R.W. Griffiths & R.I. Hill 1989. Melting in an Archaean mantle plume, heads it's basalts, tails it's komatiites. *Nature* 339: 697-699.

Campbell, R.H. 1975. *Soil slips, debris flows and rainstorms in the Santa Monica Mountains and vicinity, southern California.* US Geol. Surv. Prof. Pap. 851.

Catherall, D.J. 1973. The Shabani younger granite, Shabani, Rhodesia. *Spec. Publ. Geol. Soc. S. Africa* 3: 197-200.

Cattell, A., T.E. Krogh & N.T. Arndt 1984. Conflicting Sm-Nd whole rock and U-Pb ages for archaean lavas from Newton Township, Abitibi Belt, Ontario. *Earth Planet. Sci. Lett.* 70: 280-290.

Cawthorn, R.G. & J.R. McIver 1977. Nickel in komatiites. *Nature* 266: 716-718.

Chadwick, B., M. Romakrishnan, M.N. Viswanatha & V.S. Marthy 1978. Structural Studies in the Archaean Sargur and Dharwar Supracrustal Rocks of the Karnataka Craton. *Jour. Geol. Soc. India* 19: 531-542.

Chauvel, C., B. Dupré & G.A. Jenner 1985. The Sm-Nd age of Kambalda volcanics is 500 Ma too old. *Earth Planet. Sci. Lett.* 74: 315-321.

Chauvel, C., B. Dupré, W. Todt, N.T. Arndt & A.W. Hoffman 1983. Pb and Nd isotopic correlation in Archaean and Proterozoic greenstone belts. *EOS* 64: 330.

Christensen, U. 1985. Thermal evolution models for the Earth. *J. Geophys. Res.* 90B: 2995-3007.

Clifford, T.N. 1970. The structural framework of Africa. In T.N. Clifford & I.G. Gass (eds), *African Magmatism and Tectonics*. Edinburgh: Oliver & Boyd, 1-26.

Compston, W., I.S. Williams, I.H. Campbell & J.J. Gresham 1985. Zircon xenocrysts from the Kambalda volcanics, age constraints and direct evidence for older continental crust below the Kambalda-Norseman greenstones. *Earth Planet. Sci. Lett.* 76: 299-311.

Cotterill, P. 1969. The chromite deposits of Selukwe, Rhodesia. *Econ. Geol. Monogr.* 4: 154-186.

Coward, M.P., R.P.R. James & L. Wright 1976. Northern margin of the Limpopo Mobile Belt, southern Africa. *Geol. Soc. Am. Bull.* 87: 601-611.

Cox, K.G. 1989. Hot plumes from the mantle. *Nature* 340: 541-542.

de Wit, M.J. & R. Hart 1986. A mid-Archaean ophiolite complex, Barberton Mountain Land. Abstract. *Workshop on the tectonic evolution of greenstone belts*. Lunar and Planetary Institute, Houston. Contribution 584: 27-29.

Dimroth, E., J.A. Donaldson & J. Veizer 1981. Early Precambrian volcanology and sedimentology in the light of the recent. *Precamb. Res.* 2: 1-470.

Dodson, M.H., W. Compston, I.S. Williams & J.F. Wilson 1988. A search for ancient detrital zircons in Zimbabwean sediments. *J. Geol. Soc. Lond.* 145: 977-983.

Dupré, B. & N.T. Arndt 1986. Pb isotopic compositions of Archaean komatiites and sulfides. *Chem. Geol.* 85: 35-56.

Dupré, B. & L.M. Echeverria 1984. Pb isotopes of Gorgona Island (Colombia): Isotopic variations correlated with magma type. *Earth Planet. Sci. Lett.* 67: 186-190.

Durney, D.W. 1972. A major unconformity in the Archaean Jones Creek, Western Australia. *J. Geol. Soc. Aust.* 19: 251-259.

Echeverria, L.M. 1982. Komatiites from Gorgona Island, Colombia. In N.T. Arndt & E.G. Nisbet (eds), *Komatiites*. London: Allen & Unwin, 199-209.

Elthon, D. 1986. Komatiite genesis in the Archaean mantle, with implications for the tectonics of Archaean greenstone belts. Abstract. *Workshop on the tectonic evolution of greenstone belts*. Lunar and Planetary Institute, Houston. Contribution 584: 36-38.

Enos, P. 1977. Flow regimes in debris flows. *Sedimentology* 24: 133-142.

Ermanovics, I.F. & W.L. Davison 1976. The Pikwitonei granulites in relation to North Western Superior Province of the Canadian Shield. In B.F. Windley (ed.), *Early History of the Earth*. London: Wiley, 331-350.

Eskola, P.E. 1949. The problem of mantled gneiss domes. *J. Geol. Soc. Lond.* 104: 461-476.

Fletcher, R.A. & W.M. Espin 1897. *Geological sketch map of Matabeleland*. London: Stanfords.

Fowler, C.M.R. 1990. *The Solid Earth*. Cambridge: Cambridge University Press, 472 pp.

Galer, S.J.G., S.L. Goldstein & R.K. O'Nions 1989. Limits on chemical and convective isolation in the Earth's interior. *Chem. Geol.* 75: 257-290.

Gibbs, A.K. & W.J. Olszewski 1982. Zircon U-Pb ages of Guyana greenstone-gneiss terrane. *Precamb. Res.* 17: 199-244.

Glikson, A.Y. 1979. Early Precambrian tonalite-trondhjemite sialic nuclei. *Earth Sci. Rev.* 15: 1-73.

Goodwin, A.M. & R.H. Ridler 1970. The Abitibi orogenic belt. *Geol. Surv. Can. Pap.* 70-40: 1-24.

Groves, D.I. & W.D. Batt 1984. Spatial and temporal variations in Archaean metallogenic associations in terms of evolution of granitoid-greenstone terrains with particular emphasis on the Western Australian Shield. In A. Kroner, G.N. Hanson & A.M. Goodwin (eds), *Archaean Geochemistry*. Berlin: Springer, 73-98.

Haidutov, I.S. 1976. A greenstone belt basement relationship in the Tanganyika Shield. *Geol. Mag.* 113: 53-60.

Hamilton, P.J. 1977. Sr isotope and trace element studies of the Great Dyke and Bushveld mafic phase, and their relation to early Proterozoic magma genesis in southern Africa. *J. Petrol.* 18: 24-52.

Hamilton, P.J., R.K. O'Nions & N.M. Evensen 1977. Sm-Nd dating of Archaean basic and ultrabasic volcanic rocks. *Earth Planet. Sci. Lett.* 36: 263-268.

Hampton, M.A. 1972. The role of subaqueous debris flow in generating turbidity currents. *J. Sed. Petrol.* 42: 775-793.

Hansen, E. 1971. *Strain Facies*. Berlin: Springer, 207 pp.

Harrison, N.M. 1969. The geology around Fort Rixon and Shangani. *S. Rhodesia Geol. Surv. Bull.* 61.

227

Hawkesworth, C.J. & M.J. Bickle 1977. Rhodesian Rb-Sr geochronology from 3.6-2.0 Ga. – a brief review. *Res. Inst. Afr. Geol. Leeds Ann. Rept.* 20: 22-27.

Hawkesworth, C.J., M.J. Bickle, A.R. Gledhill, J.F. Wilson & J.L. Orpen 1979. A 2.9 b.y. event in the Rhodesian Archaean. *Earth Planet. Sci. Lett.* 43: 285-297.

Hawkesworth, C.J., S. Moorbath, R.K. O'Nions & J.F. Wilson 1975. Age relationship between greenstone belts and granites in the Rhodesian Archaean craton. *Earth Planet. Sci. Lett.* 25: 251-262.

Hawkesworth, C.J., M.J. Norry, J.C. Roddick & R. Vollmer 1979. ^{143}Nd/^{144}Nd and ^{87}Sr/^{86}Sr ratios from the Azores and their significance in LIL-element enriched mantle. *Nature* 280: 28-31.

Hawkesworth, C.J. & R.K. O'Nions 1977. The petrogenesis of some Archaean volcanic rocks from Southern Africa. *J. Petrol.* 18: 487-519.

Hayes, J.M., I.R. Kaplan & K.W. Wedeking 1983. Precambrian organic geochemistry, preservation of the record. In J.W. Schopf (ed.), *Earliest Biosphere.* Princeton University Press.

Hegner, E., A. Kroner & A.W. Hofmann 1984. Age and isotope geochemistry of the Archaean Pongola and Usushwana suites in Swaziland, southern Africa: A case for crustal contamination of mantle-derived magma. *Earth Planet. Sci. Lett.* 70: 267-279.

Hegner, E., T.K. Kyser & E.G. Nisbet 1987. Chemical and Sr-Nd-O isotopic compositions of a komatiitic lava flow from the 2.7 Ga Belingwe Greenstone Belt. Abstract. *Joint Annual Meeting, GAC MAC, Saskatoon* 12.

Henderson, J.B. 1975. Sedimentological studies of the Yellowknife Supergroup in the Slave Structural Province. *Geol. Surv. Can. Pap.* 75-1: 325-329.

Herzberg, C., T. Gasparik & H. Sawamoto 1990. Origin of mantle peridotite, constraints from melting experiments to 16.5 GPa. *J. Geophys. Res.* 95: 15779-15803.

Herzberg, C.T. & M.J. O'Hara 1985. Origin of mantle peridotite and komatiite by partial melting. *Geophys. Res. Lett.* 12: 541-544.

Hickman, M.H. 1974. 3500 m.y. old granite in southern Africa. *Nature* 251: 295-296.

Hickman, M.H. 1978. Isotopic evidence for crustal re-working in the Rhodesian Archaean craton, southern Africa. *Geology* 6: 214-216.

Hickman, M.H. & J. Wakefield 1975. Tectonic interpretation of new geochronologic data from the Limpopo Belt at Pikwe, Botswana, southern Africa. *Geol. Soc. Am. Bull.* 86: 1468-1472.

Holland, T.H. 1902. Notes on rock specimens collected by Dr. F.H. Hatchin from the Kolar Gold Field. *India Geol. Surv. Mem.* 33.

Huppert, H.E. & R.S.J. Sparks 1985a. Komatiites I, eruption and flow. *J. Petrol.* 26: 694-725.

Huppert, H.E. & R.S.J. Sparks 1985b. Cooling and contamination of mafic and ultramafic magmas during ascent through continental crust. *Earth Planet. Sci. Lett.* 74: 371-386.

Huppert, H.E., R.S.J. Sparks, J.S. Turner & N.T. Arndt 1984. Emplacement and cooling of komatiitic lavas. *Nature* 304: 19-22.

Jackson, J. & D.P. McKenzie 1983. The geometrical evolution of normal fault systems. *J. Structural Geol.* 5: 471-482.

Jahn, B.-M. & K.C. Condie 1976. On the age of the Rhodesian greenstone belts. *Contrib. Mineral. Petrol.* 57: 317-330.

James, H.L. 1951. Iron formation and associated rocks in the Iron River district, Michigan. *Bull. Geol. Soc. Am.* 62: 251-266.

Johnson, A.M. 1970. *Physical processes in geology.* San Francisco: Freeman, 577 pp.

Joliffe, A.W. 1966. Stratigraphy of the Steep Rock Group, Steep Rock Lake, Ontario. In A.M. Goodwin (ed.), *Precambrian Symposium.* Geol Assoc. Spec. Pap.: 75-98.

Keays, R.R., R. Ganapathy, J.C. Laul, U. Krahenbuhl & J.W. Morgan 1974. The simultaneous determination of 20 trace elements in terrestrial lunar and meteoritic material by radiochemical neutron activation analysis. *Analyt. Chim. Acta* 72: 1-29.

Keep, F.E. 1929. The geology of the Shabani Mineral Belt, Belingwe district. *S. Rhodesia Geol. Surv. Bull.* 12.

Kurtz, D.D. & J.B. Anderson 1979. Recognition and sedimentologic description of recent debris flow deposits from the Ross and Weddell Seas, Antarctica. *J. Sed. Petrol.* 49: 1159-1170.

Kyser, T.K. & E.G. Nisbet 1986. Stable isotopic composition of Archaean komatiites. *Terra Cognita* 6: 185.

Laubscher, D.H. 1963. The origin and occurrence of chrysotile asbestos and associated rocks in the Shabani and Mashaba areas. Ph.D. Thesis, Univ. Witwatersrand, Johannesburg.

228

Lawson, D.E. 1976. Sandstone-boulder conglomerates and a Torridonian cliffed shoreline between Gairloch and Stoer, Northwest Scotland. *Scottish J. Geology* 12: 67-88.

Lowe, D.R. & L.P. Knauth 1977. Sedimentology of the Onverwacht Group (3.4 billion years), Transvaal, South Africa, and its bearing on the characteristics and evolution of the early earth. *J. Geol.* 85: 699-723.

Macgregor, A.M. 1947. An outline of the geological history of Southern Rhodesia. *S. Rhodesia Geol. Surv. Bull.* 38.

Macgregor, A.M. 1951. Some milestones in the Precambrian of S. Rhodesia. *Proc. Geol. Soc. S. Africa* 54: 27-71.

Martin, A. 1978. The geology of the Belingwe-Shabani schist belt. *S.Rhodesia Geol. Surv. Bull.* 83.

Martin, A. 1983. The geology of the Northern part of the Belingwe Greenstone Belt and surrounding granitoids. Ph.D. Thesis, University of Rhodesia (Zimbabwe).

Martin, A., E.G. Nisbet & M.J. Bickle 1980. Archaean stromatolites of the Belingwe Greenstone Belt. *Precamb. Res.* 13: 337-362.

McKenzie, D.P. 1978. Some remarks on the development of sedimentary basins. *Earth Planet. Sci. Lett.* 40: 25-32.

McKenzie, D.P. & M.J. Bickle 1988. The volume and composition of melt generated by extension of the lithosphere. *J. Petrol.* 29: 625-679.

McKenzie, D.P., E.G. Nisbet & J.G. Sclater 1980. Sedimentary basin development in the Archaean. *Earth Planet. Sci. Lett.* 48: 35-41.

Mckenzie, D.P. & R.K. O'Nions 1991. Partial melt distributions from inversion of rare earth element concentrations. *J. Petrol.* 32: 1021-1091.

Mennell, F.P. 1904. The geology of Southern Rhodesia. *Rhodesia Museum Spec. Rept.* 2.

Mennell, F.P. 1905. Stratigraphical and Petrographical Notes on the Oldest South African Rocks. *Proc. Rhodesia Scient. Assoc.* 5: 117-133.

Menuge, J.F. 1985. Nd isotope studies of crust-mantle evolution, the Proterozoic of southern Norway and the Archaean of southern Africa. Ph.D. Thesis, University of Cambridge.

Moorbath, S., P.N. Taylor & N.W. Jones 1986. Dating the oldest terrestrial rocks: Fact and fiction. *Chem. Geol.* 57: 63-86.

Moorbath, S., P.N. Taylor, J.L. Orpen, P. Treloar & J.F. Wilson 1987. First direct dating of Archaean stromatolitic limestone. *Nature* 326: 865-867.

Moorbath, S., J.F. Wilson & P. Cotterill 1976. Early Archaean age for the Sebakwian Group at Selukwe, Rhodesia. *Nature* 264: 536-538.

Moorbath, S., J.F. Wilson, R. Goodwin & M. Humm 1977. Further Rb-Sr age and isotope data on early and late Archaean rocks from the Rhodesian craton. *Precamb. Res.* 5: 229-239.

Morrison, E.R. & J.F. Wilson 1971. Excursion Guidebook. In E.R. Morrison & J.F. Wilson (eds), *Symposium on granites, gneisses and related rocks.* Geol. Soc. S. Afr. (Rhodesian Branch).

Naldrett, A.J. & A.R. Turner 1977. The geology and petrogenesis of a greenstone belt and related nickel sulphide mineralization at Yakabindie, Western Australia. *Precamb. Res.* 3: 43-103.

Nardin, T.R., F.J. Hein, D.S. Gorsline & B.D. Edwards 1979. A review of mass movement processes, sediment and acoustic characteristics and contrasts in slope and base of slope systems versus canyon-fan-basin floor systems. *SEPM Spec. Publ.* 27: 61-73.

Nathan, H.D. & C.K. Van Kirk 1978. A model of magmatic crystallisation. *J. Petrol.* 19: 66-94.

Nesbitt, R.W. 1971. Skeletal crystal forms in the ultramafic rocks of the Yilgarn Block, Western Australia: Evidence for an Archaean ultramafic liquid. *Aust. Geol. Soc. Spec. Publ.* 3: 331-348.

Nesbitt, R.W. & S.S. Sun 1976. Geochemistry of Archaean spinifex textured peridotites and magnesian tholeiites. *Earth Planet. Sci. Lett.* 31: 433-453.

Nisbet, E.G. 1974. The geology of the Neraida area, Othris Mts., Greece. Ph.D. Thesis, Univ. of Cambridge.

Nisbet, E.G. 1982a. Definition of 'Archaean.' *Precamb. Res.* 19: 111-118.

Nisbet, E.G. 1982b. The tectonic setting of komatiites. In N.T. Arndt & E.G. Nisbet (eds), *Komatiites.* London: Allen & Unwin, 501-520.

Nisbet, E.G. 1983. Gold in the Upper Greenstones of the Belingwe Greenstone Belt. In R.P. Foster (ed.), *Gold '82.* Geol. Soc. Zimbabwe Spec. Publ. 1. Rotterdam: Balkema, 583-594.

Nisbet, E.G. 1987. *The Young Earth.* London: Allen & Unwin, 402 pp.

Nisbet, E.G., N.T. Arndt, M.J. Bickle, W.E. Cameron, C. Chauvel, M. Cheadle, E. Hegner, A. Martin,

R. Renner & E. Roedder 1987. Uniquely fresh 2.7 Ga old komatiites from the Belingwe Greenstone Belt, Zimbabwe. *Geology* 15: 1147-1150.

Nisbet, E.G. & M.J. Bickle 1982. Sedimentology of the Brooklands Formation, Zimbabwe, development of an Archaean greenstone belt in a rapidly rifted graben. Abstract. *11th Int. Congress Sediment.* IAS, Hamilton.

Nisbet, E.G., M.J. Bickle & A. Martin 1976. Some constraints on the composition of the sub-continental mantle in Archaean time. *EOS* 57: 660.

Nisbet, E.G., M.J. Bickle & A. Martin 1977. The mafic and ultramafic lavas of the Belingwe Greenstone Belt. *S.Rhodesia. J. Petrol.* 18: 521-566.

Nisbet, E.G., M.J. Bickle, A. Martin, J.L. Orpen & J.F. Wilson 1982. Komatiites in Zimbabwe. In N.T. Arndt & E.G. Nisbet (eds), *Komatiites*. London: Allen & Unwin, 97-104.

Nisbet, E.G. & G.A. Chinner 1981. Controls on the eruption of mafic and ultramafic lavas, Ruth Well Cu-Ni prospect, West Pilbara. *Econ. Geol.* 76: 1729-1735.

Nisbet, E.G., V.J. Dietrich & A. Esenwein 1979. Routine trace element determination in silicate minerals and rocks by X- ray fluorescence. *Fortschritte Mineralogie* 57: 264-279.

Nisbet, E.G., J.F. Wilson & M.J. Bickle 1981. The evolution of the Rhodesian Archaean craton and adjacent Archaean terrain. Tectonic models. In A. Kroner (ed.), *Precambrian Plate Tectonics*. Amsterdam: Elsevier, 161-181.

Norris, R.J., E.R. Oxburgh, R.A. Cliff & R.C. Wright 1971. Structural, metamorphic and geochronological studies in the Reisseck and southern Ankogel Groups. In E.R. Oxburgh (ed.), *Jahrbuch Geologischen Bundesanstalt*. Wien, 198-234.

Norrish, K. & J.T. Hutton 1969. An accurate X-ray spectographic method for the analysis of a wide range of geological samples. *Geochim. Cosmochim. Acta* 33: 431-453.

O'Hara, M.J. 1968. The bearing of phase equilibria studies on synthetic and natural systems on the origin and evolution of basic and ultrabasic rocks. *Earth Sci. Rev.* 4: 69-133.

O'Hara, M.J., M.J. Saunders & E.L.P. Mercy 1975. Garnet-peridotite, primary ultrabasic magma and eclogite: interpretation of upper mantle processes in kimberlite. *Phys. Chem. Earth* 9: 571-604.

Oldham, J.W. 1968. A short note on the recent geological mapping of the Shabani Area. In D.J.L. Visser (ed.), *Symposium on the Rhodesian Basement Complex*. Trans. Geol. Soc. S. Africa Annex, 189-194.

Oldham, J.W. 1970. A structural interpretation of the geology of the Shabani area, Rhodesia. Ph.D. Thesis, Univ. London.

Orpen, J. & J.F. Wilson 1981. Stromatolites of 3500 Ma and a greenstone-granite unconformity in the Zimbabwean Archaean. *Nature* 291: 218-220.

Orpen, J.L. 1976. Discussion of some aspects of the Archaean history of Southern Rhodesia, by M.P. Coward & R. M. Shackleton. *Res. Inst. Afr. Geol. Leeds Ann. Rept.* 20: 28-29.

Orpen, J.L. 1978. The geology of the southwestern part of the Belingwe Greenstone Belt and adjacent country – The Belingwe Peak Area. Ph.D. Thesis, Univ. Rhodesia.

Orpen, J.L., M.J. Bickle, E.G. Nisbet & A. Martin 1986. Belingwe Peak. Zimbabwe Geological Survey Map, to accompany *Geological Survey Short Report* No. 51, 1:100 000. Geological Survey Zimbabwe.

Palme, H. & E. Jagoutz 1977. Determination of major elements in small geological samples. *An. Chem.* 49: 717.

Pearce, J.A. & M.J. Norry 1979. Petrogenetic implications of Ti, Zr, Y and Nb variations in volcanic rocks. *Contrib. Mineral. Petrol.* 69: 33-47.

Prior, D.B. & E.H. Doyle 1985. Intra-slope canyon morphology and its modification by rockfall processes, U.S. Atlantic continental margin. *Marine Geol.* 67: 177-196.

Pyke, D.R., A.J. Naldrett & O.R. Eckstrand 1973. Archaean ultramafic flows in Munro Township, Ontario. *Bull. Geol. Soc. Am.* 84: 955-978.

Ramsay, J.G. 1967. *Folding and Fracturing of Rocks*. New York: McGraw-Hill, 567 pp.

Renner, R. 1989. Cooling and crystallization of komatiite flows from Zimbabwe. Ph.D. Thesis, Univ. Cambridge.

Renner, R., E.G. Nisbet, M.J. Cheadle, N.T. Arndt, M.J. Bickle & W.E. Cameron (in press). Komatiite flows from the Reliance Formation, Belingwe Belt, Zimbabwe: I – petrography and mineralogy. *J. Petrol.*

Richter, F.M. 1985. Models for the Archaean thermal regime. *Earth Planet. Sci. Lett.* 73: 350-360.

Robertson, I.D.M. 1974. Explanation of the geological map of the country south of Chibi. *Rhodesia Geol. Surv. Short Rep.* 41.

Schwerdtner, W.M. & S.B. Lumbers 1981. Major diapiric structures in the Superior and Grenville provinces of the Canadian Shield. In D.W. Strangway (ed.), *The continental crust and its mineral deposits.* Geol. Assoc. Spec. Pap. 20: 149-180.

Shirey, S.B. & G.N. Hansen 1986. Mantle heterogeneity and crustal recycling in Archaean granite-Greenstone Belts: evidence from Nd isotopes and trace elements in the Rainy Lake area, Superior Province Ontario, Canada. *Geochim. Cosmochim. Acta* 50: 2631-2651.

Smith, H.S. & A.J. Erlank 1982. Geochemistry and petrogenesis of komatiites from the Barberton Greenstone Belt, South Africa. In N.T. Arndt & E.G. Nisbet (eds), *Komatiites.* London: Allen & Unwin, 348-398.

Snowden, P.A. & M.J. Bickle 1976. The Chinamora Batholith: Diapiric intrusion or interference fold? *J. Geol. Soc. Lond.* 132: 131-137.

Stagman, J.G. 1978. An outline of the geology of Rhodesia. *Rhodesia Geol. Surv. Bull.* 80.

Steiger, R.H. & E. Jaeger 1977. Subcommission on geochemistry: Convention on the use of decay constants in geo- and cosmochronology. *Earth Planet. Sci. Lett.* 36: 359-362.

Stowe, C.R. 1968. The geology of the country south and west of Selukwe. *S. Rhodesia Geol. Surv. Bull.* 59.

Streckeisen, A.L. 1976. To each plutonic rock its proper name. *Earth Sci. Rev.* 12: 1-34.

Sun, S.S. & R.W. Nesbitt 1977. Chemical hetrogeneity of the Archaean mantle, composition derived from the chemistry of ultramafic lavas. *Nature* 263: 577-580.

Sun, S.S. & R.W. Nesbitt 1978. Petrogenesis of Archaean basic and ultrabasic volcanics: evidence from rare earth elements. *Contrib. Mineral. Petrol.* 65: 301-325.

Taylor, P.N., N.W. Jones & S. Moorbath 1984. Isotopic assesment of relative contributions from crust and mantle sources to the magma genesis of Precambrian granitoid rocks. *Phil. Trans. Roy. Soc. Lond.* A310: 605-625.

Taylor, P.N., J.D. Kramers, S. Moorbath, J.F. Wilson, J.L. Orpen & A. Martin 1991. Pb/Pb, Sm-Nd, and Rb-Sr geochronology in the Archaean craton of Zimbabwe. *Isotope Geoscience* 87: 175-196.

Taylor, S.R. & S.M. McLennan 1985. *The continental crust: Its composition and evolution.* Oxford: Blackwell, 312 pp.

Thorpe, R.I., A.H. Hickman, D.W. Davis, J.K. Mortenson & A.F. Trendall (in press). Conventional U-Pb zircon geochronology of Archaean felsic units in the Marble Bar region, Pilbara craton. *Precamb. Res.*

Viljoen, M.S. & R.P. Viljoen 1969. The geology and geochemistry of the Lower ultramafic unit of the Onverwacht Group and a proposed new class of igneous rock. *Geol. Soc. S. Africa Spec. Publ.* 2: 55-86.

Walker, D., T. Shibata & S.E. DeLong 1979. Abyssal tholeiites from the Oceanographer Fracture Zone. *Contrib. Mineral. Petrol.* 70: 111-125.

Walker, R.G. 1978. A critical appraisal of Archaean basin-craton complexes. *Can. J. Earth Sci.* 15: 1213-1218.

Watts, A.B., G.D. Karner & M.S. Steckler 1982. Lithosphere flexure and the evolution of sedimentary basins. *Phil. Trans. Roy. Soc. Lond.* A305: 249-281.

White, W.M., B. Dupré & P. Vidal 1985. Isotope and trace element geochemistry of sediments from the Barbaros Ridge-Demera Plain region, Atlantic Ocean. *Geochim. Cosmochim. Acta* 49: 1875-1886.

White, W.M. & J. Patchett 1984. Hf-Nd-Sr isotopes and incompatible element abundances in island arcs: Implications for magma origins and crust-mantle evolution. *Earth Planet. Sci. Lett.* 67: 167-185.

Wilks, M.E. & E.G. Nisbet 1985. Archaean stromatolites from the Steep Rock Group, northwestern Ontario. *Can. J. Earth Sci.* 22: 792-799.

Wilks, M.E. & E.G. Nisbet 1988. Stratigraphy of the Steep Rock Group, northwest Ontario: A major Archaean unconformity and Archaean stromatolites. *Can. J. Earth Sci.* 25: 370-391.

Williams, G.P. & M.P. Guy 1973. Erosional and depositional aspects of Hurricane Camille in Virginia, 1969. *U.S. Geol. Surv. Prof. Pap.* 804.

Wilson, J.F. 1968. The geology of the country around Mashaba. *S. Rhodesia Geol. Surv. Bull.* 62.

Wilson, J.F. 1973. The Rhodesian Archaean craton, an essay in cratonic evolution. *Phil. Trans. Roy. Soc. Lond.* A273: 389-411.

Wilson, J.F. 1979. A preliminary appraisal of the Rhodesian Basement Complex. *Geol. Soc. S. Africa Spec. Publ.* 5: 1-23.

Wilson, J.F. 1981. The granite-gneiss greenstone shield, Zimbabwe. In D.R. Hunter (ed.), *Precambrian of the Southern Hemisphere.* Amsterdam: Elsevier, 454-488.

Wilson, J.F., N. Baglow, J.L. Orpen & J.M. Tsomondo 1990. A reassessment of some regional correlations of greenstone-belt rocks in Zimbabwe and their significance in the development of the Archaean craton. Abstract. In J.E. Glover & S.E. Ho (eds), *3rd International Archaean Symposium, Perth, Geoconferences (WA) inc.,* 43.

Wilson, J.F., M.J. Bickle, C.J. Hawkesworth, A. Martin, E.G. Nisbet & J.L. Orpen 1978. Granite greenstone terrains of the Rhodesian Archaean craton. *Nature* 271: 23-27.

Wilson, J.F. & N.M. Harrison 1973. Recent K-Ar age determinations on some Rhodesian granites. *Geol. Soc. S. Africa Spec. Publ.* 3: 69-78.

Wood, R. & P. Barton 1983. Crustal thinning and subsidence in the North Sea. *Nature* 302: 134-136.

Worst, B.G. 1956. The geology of the country between Belingwe and West Nicholson. *S. Rhodesia Geol. Surv. Bull.* 43.

Yarnold, J.C. & J.P. Lombard 1989. A facies model for large rock-avalanche deposits formed in dry climates. In I.P. Colburn, P.L. Abbott & J. Minch (eds), *Conglomerates in basin analysis: A symposium dedicated to A.O. Woodford.* Pacific Section SEPM, 9-31.

York, D. 1969. Least squares fitting of a straight line with correlated errors. *Earth Planet. Sci. Lett.* 5: 320-324.

Bibliography of the Archaean crustal study project

Bickle, M.J., A. Martin & E.G. Nisbet 1975a. Basaltic and peridotitic komatiites, stromatolites and a basal unconformity in the Belingwe Greenstone Belt, Rhodesia. *Earth Planet. Sci. Lett.* 27: 155-162.

Bickle, M.J., A. Martin & E.G. Nisbet 1975b. Preliminary report on the Belingwe Greenstone Belt. *Res. Inst. Afr. Geol. Leeds Ann. Rept.* 19: 71.

Bickle, M.J., C.J. Hawkesworth, A. Martin, E.G. Nisbet & R.K. O'Nions 1976. Mantle composition derived from the chemistry of ultramafic lavas. *Nature* 263: 577-580.

Nisbet, E.G., M.J. Bickle & A. Martin 1976. Some constraints on the composition of the sub-continental mantle in Archaean time. *EOS* 57: 660.

Orpen, J.L. 1976. Discussion of some aspects of the Archaean history of Southern Rhodesia, by M.P. Coward & R. M. Shackleton. *Res. Inst. Afr. Geol. Leeds Ann. Rept.* 20: 28-29.

Bickle, M.J., C.E. Ford & E.G. Nisbet 1977. The petrogenesis of peridotitic komatiites: Evidence from high-pressure melting experiments. *Earth Planet. Sci. Lett.* 37: 97-106.

Bickle, M.J. & E.G. Nisbet 1977. The chicken or the egg. *Geology* 5: 68.

Hawkesworth, C.J. & M.J. Bickle 1977. Rhodesian Rb-Sr geochronology from 3.6- 2.0 Ga. – a brief review. *Res. Inst. Afr. Geol. Leeds Ann. Rept.* 20: 22-27.

Nisbet, E.G., M.J. Bickle & A. Martin 1977. The mafic and ultramafic lavas of the Belingwe Greenstone Belt. *S. Rhodesia. J. Petrol.* 18: 521-566.

Bickle, M.J. 1978. Heat loss from the earth: A constraint on Archaean tectonics from the relation between geothermal gradients and the rate of plate production. *Earth Planet. Sci. Lett.* 40: 301-315.

Martin, A. 1978. The geology of the Belingwe-Shabani schist belt. *S. Rhodesia Geol. Surv. Bull.* 83.

Orpen, J.L. 1978. The geology of the southwestern part of the Belingwe Greenstone Belt and adjacent country – The Belingwe Peak Area. Ph.D. Thesis, Univ. Rhodesia (Zimbabwe).

Wilson, J.F., M.J. Bickle, C.J. Hawkesworth, A. Martin, E.G. Nisbet & J.L. Orpen 1978. Granite greenstone terrains of the Rhodesian Archaean craton. *Nature* 271: 23-27.

Hawkesworth, C.J., M.J. Bickle, A.R. Gledhill, J.F. Wilson & J.L. Orpen 1979. A 2.9 b.y. event in the Rhodesian Archaean. *Earth Planet. Sci. Lett.* 43: 285-297.

Martin, A., E.G. Nisbet & M.J. Bickle 1980. Archaean stromatolites of the Belingwe Greenstone Belt. *Precamb. Res.* 13: 337-362.

McKenzie, D.P., E.G. Nisbet & J.G. Sclater 1980. Sedimentary basin development in the Archaean. *Earth Planet. Sci. Lett.* 48: 35-41.

Nisbet, E.G., J.F. Wilson & M.J. Bickle 1981. The evolution of the Rhodesian Archaean craton and adjacent Archaean terrain. Tectonic models. In A. Kroner (ed.), *Precambrian Plate Tectonics*. Amsterdam: Elsevier, 161-181.

Bickle, M.J. & K.A. Eriksson 1982. Evolution and subsidence of early Precambrian sedimentary basins. *Phil. Trans. Roy. Soc. Lond.* A305: 225-247.

Cameron, W.E. & E.G. Nisbet 1982. Possible Phanerozoic analogues of komatiitic basalts and komatiites. In N.T. Arndt & E.G. Nisbet (eds), *Komatiites*. London: Allen & Unwin, 29-50.

Nisbet, E.G. 1982a. Definition of 'Archaean.' *Precamb. Res.* 19: 111-118.

Nisbet, E.G. 1982b. The tectonic setting of komatiites. In N.T. Arndt & E.G. Nisbet (eds), *Komatiites*. London: Allen & Unwin, 501-520.

Nisbet, E.G. & M.J. Bickle 1982. Sedimentology of the Brooklands Formation, Zimbabwe, development of an Archaean greenstone belt in a rapidly rifted graben. Abstract. *11th Int. Congress Sediment.* IAS, Hamilton.

Nisbet, E.G., M.J. Bickle, A. Martin, J.L. Orpen & J.F. Wilson 1982. Komatiites in Zimbabwe. In N.T. Arndt & E.G. Nisbet (eds), *Komatiites*. London: Allen & Unwin, 97-104.

Martin, A. 1983. The geology of the Northern part of the Belingwe Greenstone Belt and surrounding granitoids. Ph.D. Thesis, University of Rhodesia (Zimbabwe).

Nisbet, E.G. 1983. Gold in the Upper Greenstones of the Belingwe Greenstone Belt. In R.P. Foster (ed.) *Gold '82*. Geol. Soc. Zimbabwe Spec. Publ. 1. Rotterdam: Balkema, 583-594.

Abell, P.I., J. McClory, A. Martin & E.G. Nisbet 1985b. Archaean stromatolites from the Ngesi Group, Belingwe Greenstone Belt, Zimbabwe; preservation and stable isotopes-preliminary results. *Precamb. Res.* 27: 357-383.

Abell, P.I., A. McClory, A. Martin, E.G. Nisbet & T.K. Kyser 1985a. Petrography and stable isotope ratios from Archaean stromatolites, Mushandike Formation, Zimbabwe. *Precamb. Res.* 27: 385-398.

Bickle, M.J. 1986. Implications of melting for stabilisation of the lithosphere and heat loss in the Archaean. *Earth Planet. Sci. Lett.* 80: 314-324.

Kyser, T.K. & E.G. Nisbet 1986. Stable isotopic composition of Archaean komatiites. *Terra Cognita* 6: 185.

Orpen, J.L., M.J. Bickle, E.G. Nisbet & A. Martin 1986. Belingwe Peak. Zimbabwe Geological Survey Map, to accompany *Geological Survey Short Report* No. 51, 1:100 000. Geological Survey Zimbabwe.

Hegner, E., T.K. Kyser & E.G. Nisbet 1987. Chemical and Sr-Nd-O isotopic compositions of a komatiitic lava flow from the 2.7 Ga Belingwe Greenstone Belt. Abstract. *Joint Annual Meeting, GAC MAC, Saskatoon* 12: 55.

Nisbet, E.G. 1987. *The Young Earth*. London: Allen & Unwin, 402 pp.

Nisbet, E.G., N.T. Arndt, M.J. Bickle, W.E. Cameron, C. Chauvel, M. Cheadle, E. Hegner, A. Martin, R. Renner & E. Roedder 1987. Uniquely fresh 2.7 Ga old komatiites from the Belingwe Greenstone Belt, Zimbabwe. *Geology* 15: 1147-1150.

Renner, R. 1989. Cooling and crystallization of komatiite flows from Zimbabwe. Ph.D. Thesis, Univ. Cambridge.

Renner, R., E.G. Nisbet, M.J. Cheadle, N.T. Arndt, M.J. Bickle & W.E. Cameron (in press). Komatiite flows from the Reliance Formation, Belingwe Belt, Zimbabwe: I – Petrography and mineralogy. *J.Petrol.*

Subject index

Printed and bound by CPI Group (UK) Ltd, Croydon, CR0 4YY

23/10/2024

01777686-0017